高等职业教育畜牧兽医类专业教材

畜牧兽医专业英语

刘　宁　主编

中国轻工业出版社

图书在版编目（CIP）数据

畜牧兽医专业英语/刘宁主编．—北京：中国轻工业出版社，2021.1

ISBN 978-7-5184-2801-4

Ⅰ.①畜…　Ⅱ.①刘…　Ⅲ.①畜牧学-英语-高等职业教育-教材 ②兽医学-英语-高等职业教育-教材　Ⅳ.①S81 ②S85

中国版本图书馆 CIP 数据核字（2020）第 181460 号

责任编辑：贾　磊　王昱茜　　责任终审：张乃柬　　封面设计：锋尚设计
版式设计：砚祥志远　　责任校对：方　敏　　责任监印：张　可

出版发行：中国轻工业出版社（北京东长安街 6 号，邮编：100740）
印　　刷：北京君升印刷有限公司
经　　销：各地新华书店
版　　次：2021 年 1 月第 1 版第 1 次印刷
开　　本：720×1000　1/16　印张：19
字　　数：370 千字
书　　号：ISBN 978-7-5184-2801-4　定价：46.00 元
邮购电话：010-65241695
发行电话：010-85119835　传真：85113293
网　　址：http：//www.chlip.com.cn
Email：club@chlip.com.cn
如发现图书残缺请与我社邮购联系调换
180257J2X101ZBW

本书编写人员

主　编　刘　宁（河南科技大学）

副主编　王建平（河南科技大学）
　　　　薛琳琳（黑龙江职业学院）

参　编　（按姓名拼音排序）
　　　　陈玉洁（内蒙古农业大学职业技术学院）
　　　　汤　莉（信阳农林学院）
　　　　赵存真（信阳农林学院）
　　　　赵云翔（佛山科学技术学院）

前 言 PREFACE

畜牧兽医专业英语是高等职业院校畜牧兽医类专业学生的重要基础课。近年来，随着我国进一步推行改革开放，外资企业在我国投资建厂持续增加，我国也在不断加大对外贸易活动。越来越多的公司为了拓展公司业务，需要引进专业英语水平高的人才。这就对畜牧兽医专业英语教学提出了更高的要求，必须立足现实情况，改变现行落后的英语教学模式，开展以培养学生职业能力和社会需求相适应的英语教学，切实提高学生的英语实际应用水平和职业素养。而英语教学的改革首先是教材的改革，必须要有与目前需求相适应的教材。为此我们组织相关院校既具有较好畜牧兽医专业知识，又有丰富专业英语教学经验的一线教师编写了《畜牧兽医专业英语》教材。

在编写过程中，我们充分把英语与畜牧兽医专业相结合，努力实现以培养学生综合能力为目的的专业英语教学，既注重英语语法等基础知识的传授，更重视培养学生的实际应用能力；在教学素材选用方面既充分考虑畜牧兽医类专业特点，又重视英语学习的规律。本教材所有选材均为英语文献原文，旨在让学生学习纯正的英文。在课程内容设计方面提供了大量专业释义，并提供了课文参考译文，帮助学生理解原文，提高语言模仿效果，掌握高效学习方法。总目标是让学生通过专业英语学习，可以在将来的工作中轻松地运用所学知识看懂相关专业英文。

畜牧兽医专业内容丰富，限于篇幅和教学时数，不可能做到面面俱到。本教材在编写过程中考虑到畜种差异和知识结构的系统性，力争做到点和面相互照应，在教学过程中可以结合培养目标进行节选。

本教材由刘宁担任主编、王建平统稿。具体编写分工：赵云翔编写第Ⅰ单元第1~4课和第Ⅳ单元的第16课；陈玉洁编写第Ⅱ单元第5~9课；刘宁编写第Ⅲ单元第10~13课；汤莉编写第Ⅳ单元第18~19课；赵存真编写第Ⅳ单元第14、15、17、20课；薛琳琳编写第Ⅴ单元第21~24课；王建平编写附录。

本教材的学习对象以高等职业院校畜牧兽医类专业学生为主，其他相关专业及大专、中专师生及畜牧兽医相关科研人员、生产技术人员和畜牧兽医爱好者也可参考学习。

本教材在编写过程得到了美国佐治亚大学王锦泉博士的大力帮助，在此深表感谢。教材中参考和引用了大量文献资料，在此对原作者一并表示感谢！

由于编者水平所限，不妥之处在所难免，殷切希望专家学者提出批评和建议，以便再版时修改和完善。

编者

2020年9月

Contents 目录

Unit Ⅰ Breeding

Unit Ⅱ Nutrition and Feeding

Unit Ⅲ Management

Unit Ⅳ Animals Care

目 录 Contents

第Ⅰ单元 育种

第Ⅱ单元 营养与饲养

第Ⅲ单元　管理

第Ⅳ单元　动物护理

第Ⅴ单元　健康

附录

Unit Ⅰ Breeding

Lesson 1
Genetics

Part One Intensive Reading

1. Learning Objective

After learning this lesson, you should understand the following:

(1) Define the concept of genetic manipulation of variance of phenotype.

(2) Give the reason why the meat processors often desire a uniform product.

(3) Give an overview of why we can not achieve genetic uniformity by high levels of inbreeding.

(4) Explain why selection for uniformity is unlikely to be successful.

(5) Describe what difficulties may be caused by genes in effects on productivity.

(6) Explain why the optimum management for animals with one genotype may not be optimum for other genotypes.

2. Text A Genetic Manipulation of Variance of Phenotype

Variance of phenotype in commercial production can have direct effects on profitability. Many production systems would be more profitable with increased uniformity of phenotype. Large variation in phenotype may mean that a final management system is suboptimal for all but a few animals in the system. Similarly, meat processors often desire a uniform product to optimize slaughter operations and retailing of final product. It is also possible that increased phenotypic variance would be profitable in some cases. For example, with dairy cattle, increased phenotypic variance at a fixed mean of performance of cows entering the herd would lead to a greater increase in herd mean production level (across all age classes) for a given rate of voluntary culling. The increase in herd profitability would presumably be dependent on the rate of voluntary culling and rate of change in production with age and distribution of age classes in the herd.

Genetic uniformity can be achieved by high levels of inbreeding, but this inevitably leads to reduced performance (inbreeding depression) and reduced homeostatic control and consequent increased phenotypic variance. Under an infinitesimal model, selection is expected to have little effect on genetic variation (except for the temporary effect of gametic phase disequilibrium). And, selection for uniformity is unlikely to be successful since, (1) homeostasis is expected to be highly related to fitness and hence show little or no genetic variation in the positive direction (*e. g.* increased homeostasis is equal to reduced phenotypic variance), and (2) homeostasis is associated with heterozygosity in naturally outbreeding species. Thus, on our present understanding, breeding programs to reduce phenotypic variance of quantitatively inherited traits are unlikely to be successful. And, programs to increase phenotypic variance, via inbreeding, would likely pay an unacceptable cost of inbreeding depression of average performance.

Increased uniformity can be achieved for traits controlled by few genes, by fixation of the desirable alleles. A few examples are the coat and skin color patterns of recognized breeds, the halothane gene in pig (causing increased stress susceptibility and carcass leanness), the Booroola gene in sheep (causing large increases in lambing rates), the double muscling gene in beef cattle (causing substantial increases in carcass lean weight and percentage) and the dwarfing gene in poultry (used in female parents of broiler stocks to reduce body size and hence egg production costs).

Segregation of genes determining coat and skin color patterns can cause the visual perception of large amounts of phenotypic variation while, in many production systems, having no effect on variation in productivity. In some situations such variation may however be associated with variation in productivity via effects on susceptibility to such productivity reducing complaints as sunburn (pale skinned pigs), eye cancer (cattle with lightly pigmented eye surrounds) and tick burden (dark coated cattle). And, even where color pattern variation has no direct effect on productivity, it may have substantial effects on profitability of a breeding company since customers often expect or prefer a particular color pattern, even to the extent of refusing to purchase a particular stock if of an unusual or unexpected coloration. In such cases it is important not just to have the optimum color pattern on average, but to have as little variation as possible about that optimum pattern.

Similarly, genes which have large effects on productivity may cause management difficulties if segregating in a population since optimum management for animals of one genotype may not be optimum for other genotypes. An example might be the Booroola gene. The effect of the Booroola gene varies depending on genetic background, but a typical situation might be average litter sizes of 1, 2 and 3+ for the homozygous wild type, the heterozygotes and homozygous booroola. Under harsh extensive conditions, a litter size of 1 is optimum, leading to elimination of booroola allele. Under semi-intensive conditions, a litter size of 2 could be optimum for a dam line suggesting a crossing scheme between two lines fixed for opposite alleles to produce F_1 dams. Under highly intensive conditions a litter size of 3 or more might be optimum leading to fixation of the Booroola allele. Perhaps more likely, a proportion of litters of 3+ might be acceptable provided there were also single litters for fostering of lambs from ewes unable to cope properly with 3 or more lambs. In such situations a segregating population might be desirable.

3. New Words and Phrases

breed [briːd] *vi.* 繁殖；饲养；产生；*vt.* 繁殖；饲养；养育，教育；引起；*n.* [生物] 品种；种类，类型

genetic manipulation 遗传调控

commercial [kəˈmɜːrʃl] *adj.* 商业的；贸易的；以获利为目的的

profitability [ˌprɑfɪtəˈbɪlətɪ] *n.* 获利（状况），盈利（情况）

uniformity [ˌjuːnɪˈfɔːmətɪ] *n.* 一致性；均匀性

suboptimal [ˈsʌbˈɒptɪməl] *adj.* 未达最佳标准的；非最理想的

slaughter [ˈslɔːtə] *vt.* 屠宰，屠杀；杀戮；使惨败；*n.* 屠宰，屠杀；杀戮；消灭

voluntary culling 主动淘汰

presumably [prɪˈzuːməbli] *adv.* 大概；可能

polygenic [ˌpɑlɪˈdʒenɪk] *adj.* 多基因的（遗传特征）

inevitably [ɪnˈevɪtəbli] *adv.* 难免；不可避免地

homeostatic [ˌhoʊmɪˈsteɪtɪk] *adj.*（社会群体的）自我平衡的，原状稳定的
hence [hɛns] *adv.* 从此；因此，所以
inbreeding [ˈɪnbriːdɪŋ] *n.* 近亲交配；同系繁殖
depression [dɪˈpreʃən] *n.* 萎靡不振，沮丧；下陷处，坑；衰弱；减缓
susceptibility [səˌseptəˈbɪlɪti] *n.* 易受影响或损害的状态，易感性
perception [pərˈsepʃən] *n.* 知觉；观念
be associated with 与……有关
complaints [kəmˈpleɪnts] *n.* 投诉；抱怨
purchase [ˈpɜːrtʃəs] *v.* 购买
optimum [ˈɑːptɪməm] *adj.* 最适宜的
homozygous [ˌhɒməˈzaɪɡəs] *n.* 纯合子
heterozygote [ˌhetərəˈzaɪɡoʊt] *n.* 杂合子
extensive [ɪkˈstɛnsɪv] *adj.* 广阔的，广大的
intensive [ɪnˈtɛnsɪv] *adj.* 加强的，强烈的
scheme [skim] *n.* 计划
ewe [juː] *n.* 母羊

4. Notes to the Text A

（1）Large variation in phenotype may mean that a final management system is suboptimal for all but a few animals in the system.

如果出现了大量的表型变异，说明这个群体的性能不稳定，比如产仔数可能会忽高忽低，而优秀的性能也不能稳定遗传给下一代，从而不能创造良好的经济效益。"mean that"后面为宾语从句。

（2）Similarly, meat processors often desire a uniform product to optimize slaughter operations and retailing of final product.

请思考，是整齐划一的商品更好加工还是形态各异的商品更好加工？答案显而易见。作为肉类加工者，当然是希望猪只越整齐均匀越好。这一切都需要一个稳定的表型来维持。

（3）Genetic uniformity can be achieved by high levels of inbreeding, but this inevitably leads to reduced performance (inbreeding depression) and reduced homeostatic control and consequent increased phenotypic variance.

近交从血缘关系上来讲，由于相似度高，因此可以提高遗传的一致性。但是近亲交配容易产生品质低下的后代，这是由近交衰退导致的表型变异。因此我们不可能通过近交手段来实现遗传一致性。

（4）Increased uniformity can be achieved for traits controlled by few genes, by fixation of the desirable alleles.

有一些性状只由少量的基因调控，因此可以通过生物学手段对这些基因进行控制，从而增加性状的一致性。但对于由很多基因共同调控的性状，操作起来相对复杂。

（5） Similarly, genes of large effects on productivity may cause management difficulties if segregating in a population since optimum management for animals of one genotype may not be optimum for other genotypes.

由于主效基因的影响，基因分离会导致群里变异度加大，造成生产管理的难度增加，因为遗传育种工作对优化畜群的生产管理有一定的影响。

句中“since”为原因状语从句，后面的句子是对前面的“management difficulties if segregating in a population”解释和补充。

5. Exercises

（1） Fill in the blanks by finishing the sentences according to the passage.

①The increase in herd profitability would presumably be dependent on the rate of ________ and ________ of change in production with age and distribution of age classes in the herd.

②________ can be achieved by high levels of inbreeding, but this inevitably leads to reduced performance (inbreeding depression) and reduced homeostatic control and consequent increased phenotypic variance.

③Under an infinitesimal model, ________ is expected to have little effect on genetic variation (except for the temporary effect of gametic phase disequilibrium).

④Segregation of genes determining coat and skin color patterns can cause the visual perception of large amounts of ________ while, in many production systems, having no effect on variation in productivity.

⑤Similarly, genes of large effects on productivity may cause ________ if segregating in a population since optimum management for animals of one genotype may not be optimum for other genotypes.

⑥In such situations a ________ might be desirable.

（2） Answer the following questions according to the passage.

①What does the genetic manipulation of variance of phenotype mean?

②Why do the meat processors often desire a uniform product?

③Is it possible that increased phenotypic variance would be profitable in some cases?

④Can we achieve genetic uniformity by high levels of inbreeding?

⑤Why selection for uniformity is unlikely to be successful?

⑥What difficulties may cause in genes of large effects on productivity?

(3) Translation of the following sentences into Chinese.

①Variance of phenotype in commercial production can have direct effects on profitability. Many production systems would be more profitable with increased uniformity of phenotype.

② Similarly, meat processors often desire a uniform product to optimize slaughter operations and retailing of final product.

③It is also possible that increased phenotypic variance would be profitable in some cases.

④ Homeostasis is associated with heterozygosity in naturally outbreeding species.

⑤ And, programs to increase phenotypic variance, via inbreeding, would likely pay an unacceptable cost of inbreeding depression of average performance.

⑥Increased uniformity can be achieved for traits controlled by few genes, by fixation of the desirable alleles.

课文A　表型变异的遗传学调控

商业生产中表型变异对经济效益有直接影响。随着表型均匀性的增加，许多生产系统将变得更有收益。表型的巨大变化可能意味着，除了系统中的少数动物外，最终管理系统对于所有动物来说都不够理想。同样，肉类加工者往往希望得到一个统一的产品来优化屠宰业务和最终产品的零售。在某些情况下，增加表型变异也可能是有益的。例如，对于奶牛，进入牛群的奶牛在一定的性能水平上（所有年龄段）的表型差异增加，将导致在一定程度上期望宰杀的牛群平均生产水平的增加。牧群盈利能力的增加大概取决于在牧群中随年龄和年龄分布而产生的期望淘汰率和产量变化率。

遗传的一致性可以通过高水平的近亲繁殖来实现，但这不可避免地导致性能下降（近交衰退），并减少稳态控制，从而增加表型变异。在一个无限小的模型中，选择对遗传变异的影响不大（除了配子相不平衡的暂时效应）。选择一致性不太可能成功的原因为；（1）稳态被认为与健康密切相关，因此在正方向上很少或没有遗传变异（即增加稳态意味着表型变异减少）。(2）稳态与自然繁殖物种的杂合性有关。因此，以我们目前的理解，近亲繁殖的育种计划减少表型变异后，在定量遗传特征方面不大可能成功。

而且，通过近亲繁殖来增加表型变异的方案，可能会因近交衰退给平均生产性能带来不可接受的代价。

通过对理想等位基因的控制，可以实现少数基因控制的性状的一致性增强。例如，某些品种的皮毛和皮肤颜色控制的基因、猪的氟烷基因（对压力敏感性增加和胴体异常）、绵羊的多胎基因（可以增加多胎性）、肉牛的双肌基因（导致胴体质量和百分比大幅增加）、家禽中的矮化基因（可使母鸡的体型减小，从而降低鸡蛋的生产成本）。

在许多生产系统中，决定皮毛和皮肤颜色图案的基因分离会引起大量表型变异在视觉上的变化，而对生产力的变化没有影响。然而，在某些情况下，这种变异可能会对生产力产生影响，它会减少遗传病，如晒斑（皮肤苍白的猪）、眼癌（眼睛周围有色素沉着的牛）和感染寄生虫（深色牛）。即使颜色模式变化没有直接影响生产率，它也可能对经济效益产生实质性的影响，育种公司因为客户往往希望或喜欢一个特定的颜色模式，甚至在某种程度上拒绝购买特殊的颜色，如果出现一个不寻常的或意想不到的颜色。在这种情况下，最重要的是，不仅要有最佳的颜色模式，而且要尽可能少地改变最佳的模式。

同样地，对产量有极大影响的基因可能会导致管理上的困难，因为一种基因型动物的最佳管理可能并不适合其他基因型。例如多胎基因，多胎基因的效果取决于遗传背景，但典型的情况可能为野生型纯合子、杂合子和隐性纯合子的平均产仔数分别为1、2和3+。在恶劣的条件下，一胎产1仔的情况是最适宜的，从而导致了隐性基因的基因频率降低。在半集约化的条件下，一胎产2仔可能是最适宜的可通过杂交方案来控制F_1代。在高强度条件下，3个或3个以上的胎仔数可能是最优的，需要显性等位基因的固定。更有可能的是，如果有产3个仔以上的羔羊，也可以被接受，它们可以被转移给产一个羔羊的母羊，在母羊无法妥善处理3只或更多的羊羔情况下。在这种情况下，人为地隔离是可取的。

Part Two Extensive Reading

1. Text B Animal Genetics and Welfare

Animal genetics have led to a remarkable increase in animal productivity. Increased production of milk, meat, eggs, and fiber has been critically needed for a growing world. Genetic engineering has provided excellent tools to develop animal models to study and find possible treatment for devastating diseases such as cancer, diabetes, cardiovascular diseases, and Alzheimer's disease. However, extensive genetic selection based on single production traits can compromise the welfare of animals. Such welfare-related problems include reduced reproductive efficiency and increased disease susceptibility in dairy cattle, skeletal disorders and behavioral change in poultry, and cardiac arrest and lameness in pigs. To decrease welfare issues in animals, breeding strategies have had to combine selection for productivity along with welfare to assure sustainable farm animal production.

Establishment of a trait assessing system using physiological, phenotypic, and behavioral indicators will improve genetic selection and speed detection of welfare and production related problems, problems that can be addressed quickly. Breeding and genetic selection is fundamentally based on economics. Increasingly welfare fits into the economics of any breeding program. With the development of new genetic and non-genetic tools such as genomic selection and genetic markers, animal productivity, health, and welfare will continue to improve and advance production goals.

A continued need for genetic improvements and knowledge

Use of modern animal breeding techniques has dramatically improved animal productivity (*e.g.*, growth, feed efficiency, reproduction, and disease resistance). For example, in the last 40 years milk production per cow has nearly doubled due to genetic selection and management strategies to realize genetic potential. Broiler carcass weight increased nearly five-fold during a similar 40-year period. Such advances have assured an affordable supply of animal products for human consumption and efficient use of feedstocks to produce these products. According to the United Nations Food and Agricultural Organization (FAO) food production will need to continue to expand (approximately 1% per year just to meet population growth) for the next few decades in order to feed the world's population; a population expected to reach 9 billion by 2070. As global economies grow, the demand for animal products will also expand (a demand on meat production that exceeds population growth). Global meat consumption has doubled since 1950 and continued expansion is predicted. In the past, increased human consumption of animal products has been complemented with increased animal productivity. In the future, expanded demand for increased animal products, particularly in emerging economies and no additional landmass for cultivation, will require accelerated genetic improvements of livestock and poultry growth, feed efficiency, and reproduction using both traditional breeding and genetic engineering.

Genetic advances in agricultural animal productivity will require new methods to manage increased genetic potential. Added productivity without improved management strategies can adversely compromise animal welfare. For example, dairy nutritionists found that the energy demands for high milk producing cows had to be managed through the use of improved feeding practices. Without improved management practices, the genetic potential was either not realized or the animal's welfare was compromised. However, all genetic manipulations do not have management fixes and some may result in chronic welfare concerns. In situations where animal welfare cannot be maintained by the caretaker, scientists may have to step forward and set limits on genetic changes that result in a chronic animal welfare concerns that are not acceptable. Failure to set scientifically based limits could result in public concern and ultimately laws that further regulate the use of animals both in agriculture and in science.

Increasingly, genetics for improved animal productivity are in the hands of fewer and fewer commercial genetics companies. The companies that supply most of the genetics for swine, laying hens, and broiler chickens are controlled by a handful of primary breeders. Competition to remain as a primary source of genetics in these animal agricultural industries is fierce. While there have been breeding programs that have adversely affected animal welfare at times, genetics companies cannot remain profitable if the overall welfare of the animal is compromised to a point where animal productivity is adversely affected. Hence, the welfare of the animal is a primary concern to all commercial genetics companies. Just as animal genetics and breeding have greatly contributed towards efficient use of land and agricultural resources leading to increase in farm production and improved food quality and security, genetic selection and engineering will be critical to continued improvement in productivity, food quality, and the welfare of the animal.

In other ways, genetically engineered animals have served as critical models for the advancement of both human and animal medicine. For example, genetically engineered mice have served as key models for the development of therapies for a number of cancers. Advancements in the creation of genetically "defective" animals have demanded the development of new methods to manage animal welfare. Mice that were severely immunocompromised could be managed in environmental conditions that minimized exposure to potentially lethal pathogens. As with farmed animals, scientist may find they have to either develop specialize management of genetically engineered models or in some cases decide that welfare is compromised to such a great extent, that the animal model should be abandoned.

2. Notes to the Text B

(1) To decrease welfare issues in animals, breeding strategies have had to combine selection for productivity along with welfare to assure sustainable farm animal production.

为了减少动物的福利问题，育种方案必须将性状选择与福利相结合，以确保可持续的农场动物生产。

(2) In the future, expanded demand for increased animal products, particularly in emerging economies and no additional landmass for cultivation, will require accelerated genetic improvements of livestock and poultry growth, feed efficiency, and reproduction using both traditional breeding and genetic engineering.

未来，对增长的动物产品的需求不断扩大，特别是在新兴经济体，没有额外的土地用于耕种，将需要利用传统育种和基因工程相结合加速遗传改良牲畜和家禽的生长、饲料效率和繁殖能力。

句中“particularly”一词表示强调。

(3) While there have been breeding programs that have adversely affected animal welfare at times, genetics companies cannot remain profitable if the overall welfare of the animal is compromised to a point where animal productivity is adversely affected.

虽然有些育种规划偶尔会对动物福利产生不利影响，如果动物的整体福利受到影响，动物产量也会受到不利影响，那么遗传公司就无法保持盈利。

3. Answer Questions

(1) What have animal genetics led to a remarkable increase in?

(2) What is breeding and genetic selection fundamentally based on?

(3) What has animal productivity dramatically been improved?

(4) What is the reason the genetic advances in agricultural animal productivity require new methods to manage increased genetic potential?

(5) Why is the welfare of the animal a primary concern to all commercial genetics companies?

课文B　动物遗传

动物遗传学的应用使畜牧业生产力显著提高。对于一个不断发展的世界来说，急需增加牛乳、肉类、蛋类和纤维的产量。基因工程为开发动物模型提供了极好的工具，可用于研究和治疗癌症、糖尿病、心血管疾病和阿尔茨海默病等破坏性疾病。然而，基于单一生产性状的广泛遗传选择可能会损害动物的福利。这些与福利相关的问题包括降低生殖能力、增加奶牛的疾病易感性、家禽的骨骼疾病和行为异常以及猪的心脏骤停和跛足。为了减少动物的福利问题，育种方案必须将性状选择与福利相结合，以确保农场动物的可持续生产。

使用生理、表型和行为指标建立福利性状的评估系统将改善福利相关性状的遗传选择和检测速度以及与生产力相关问题。育种和遗传选择基本上是基于经济学。人们越来越多地将福利性状纳入育种计划的经济权重中。随着基因组选择和遗传标记等新的遗传和非遗传工具的发展，动物生产、健康和福利将继续改善，推进生产目标。

遗传改良和知识的持续需要

使用现代动物育种技术显著提高了动物生产力（如生长、饲料效率、繁殖和抗病性）。例如，在过去40年中，由于遗传选择和管理策略实现了遗传潜力的发挥，奶牛单产量几乎翻了一番。同样地，在过去的40年间，肉鸡胴体产量几乎增加了近五倍。这些进步确保人类消费动物产品的充足供应和饲料的有效利用。根据联合国粮食及农业组织（FAO）的说法，粮食生产需要在未来几十年内继续扩大（每年约1%的增长才能满足人口增长）以养活全世界的人口。预计到2070年世界人口将达90亿。随着全球经济的增长，对动物产品的需求也将扩大（对肉类生产的需求将超过人口增长）。自1950年以来，全球肉类消费量翻了一番，预计将持续增长。在过去，人类对动物产品的消费增加与动物生产力的提高相辅相成。未来，对增长的动物产品的需求不断扩大，特别是在新兴经济体，而且没有额外的土地用于耕种，将需要利用传统育种和基因工程相结合加速遗传改良牲畜和家禽的生长、饲料效率和繁殖能力。

农业动物生产力的遗传进展需要新的方法来管理以增长遗传潜力。在没有改进管理策略的情况下增加生产力可能会降低动物福利。例如，奶牛营养学家发现，必须通过优化喂养方法来满足高产奶牛的能量需求。如果没有优化管理实践，要么遗传潜力没有实现，要么动物的福利受到损害。然而，所有遗传操作不配套管理优化，可能会导致长期的动物福利问题。在看护人无法维持动物福利的情况下，科学家可能不得不向前迈进，并对遗传变化设定限制，减少动物无法承受的慢性福利问题。未能设定科学的限制可能会引起公众关注，并最终导致针对农业动物和试验动物的规范性法律出台。

用于提高动物生产力的遗传技术被集中在少数的商业遗传公司中。猪、蛋鸡和肉鸡有关的育种公司由少数初级繁育者所控制，育种企业间关于基础种源的竞争越来越激烈。虽然有些育种规划偶尔会对动物福利产生不利影响，如果动物的整体福利受到影响，动物生产力也会受到不利影响，那么遗传公司就无法保持盈利。因此，动物的福利是所有商业遗传公司的主要关注点。动物遗传和育种技术提高了对土地和农业资源的有效利用，并且增加了农产品产量，提高了食品质量和安全，遗传选择和基因工程对于持续提高生产力、食品质量和动物福利至关重要。

在其他方面，基因工程动物已成为推动人类和动物医学发展的关键模型。例如，基因工程小鼠已成为许多癌症治疗发展的关键模型。构建遗传上“有缺陷”的动物要求开发新的方法来管理动物福利。具有严重免疫缺陷的小鼠需要最大限度地减少暴露在潜在致命病原体的环境条件。与养殖动物一样，科学家要么必须研究基因工程模型动物的专业化管理方法，要么在某些情况下模型动物的福利在很大程度上受到损害时，应该放弃动物模型的研究。

扫码进行拓展学习

Lesson 2
Breeding

Part One Intensive Reading

1. Learning Objective

After learning this lesson, you should understand the following:

(1) Describe why breeding goals and objectives should fit the strengths (or niche) of each breed and how to design to meet specific needs of targeted markets.

(2) Give an overview of crossbreeding system and production methods being used by potential customers and unique marketing opportunities should be considered in breeding goals and objectives.

(3) Define the meaning of across herd genetic ties.

(4) Describe how to properly designed a contemporary groups.

(5) Explain why breeders should plan their mating to prevent inbreeding.

2. Text A Seedstock Producers

Seedstock producers should have well defined breeding goals and objectives for each breed or line of swine that they raise. These goals should fit the strengths (or niche) of each breed and be designed to meet specific needs of targeted markets. There are opportunities for different selection objectives based on various crossbreeding programs, marketing opportunities and production methods (confinement versus outdoor) used by potential customers.

The type of crossbreeding system being used by potential customers should be considered in deciding which traits to emphasize and which breeds and breed crosses to produce. Some commercial producers mate lean, fast growing boars (terminal sires) to prolific crossbred females (maternal lines) with all resulting pigs going to market. Other producers rotate several breeds of dual purpose sires in the production of market hogs and replacement gilts. Breeds used to produce terminal sires should emphasize postweaning traits. In maternal and dual purpose breeds, producers should select on a combination of reproductive and postweaning traits.

Unique marketing opportunities should also be considered. Opportunities exist to market pork that excels in meat quality. For breeds excelling in meat quality, producers should include this characteristic in their selection program along with other important traits.

In determining selection objectives, breeders might also consider the production methods used by potential customers. Total confinement with slatted floors, hoop structures with bedded floors, and outdoor farrowing are examples of some different production methods. A breeder selling animals to herds with sows on pasture might have a slightly different selection emphasis than a breeder who sells boars and gilts to confinement operations. For customers with pasture operations, the breeder might consider temperament as an additional trait in their selection program. Temperament would be important since their customers would need docile sows with good mothering instinct and requiring minimum care. These operations would also need active boars for pen mating on pasture.

Seed stock producers need to produce the best breeding stock possible for their customers. To achieve this goal, an effective selection program is needed. Most selection programs include both within herd selection as well as selecting outside boars and semen. An effective within herd selection program must be well organized. Records should be collected on most of the herd and processed in a timely manner by a genetic evaluation program so it is possible to make meaningful comparisons. In selecting animals, meaningful comparisons are possible when breeders have properly designed contemporary groups. A properly designed contemporary group includes animals which have common sex and environment. Contemporary groups should consist of at least 20 pigs from 5 litters and 2 or more sires. Ideally, one of these sires is used by other breeders, thus resulting in across herd genetic ties.

Across herd ties are important for accurate genetic evaluations. The purchase of semen is a common way to access these reference sires and provide for genetic ties among herds. Having a reasonable size contemporary group is important for reliable genetic evaluations. Furthermore, a contemporary group should have no more than a three-to four-week span in ages to help reduce environmental differences.

Equipments and techniques that allow collection of accurate records should be used. Accuracy of performance testing is improved by utilizing real-time ultrasound technicians that are certified by the National Swine Improvement Federation. Producers should use testing methods which consists of recording all litters and performance testing at least 50% of the pigs weaned. Records should be processed in a timely manner by genetic evaluation programs. Breeders should use records in selecting the best animals to replace lower ranking sires and sows. Finally, breeders should plan their mating to prevent inbreeding.

3. New Words and Phrases

define [dɪˈfaɪn] *v.* 规定；使明确；（给词、短语等）下定义
niche [niːʃ] *n.* 合适的位置（工作等）；有利可图的缺口，商机
opportunity [ˌɑːpərˈtuːnəti] *n.* 机会
confinement [kənˈfaɪnmənt] *n.* 关押；分娩；限制，约束
versus [ˈvɜːrsəs] *prep.* 对抗；（比较两种想法、选择等）与……相对
potential [pəˈtɛnʃəl] *adj.* 潜在的
emphasize [ˈɛmfəˌsaɪz] *v.* 强调，着重；使突出
rotate [ˈroʊteɪt] *v.* 使转动；使轮流，轮换；交替
hog [hog] *n.* 育肥猪
replacement [rɪˈplesmənt] *n.* 替换；更换；替代品
terminal [ˈtɜːrmɪnl] *adj.* 末端的；末期的
excel [ɪkˈsɛl] *v.* 优于，擅长
alongwith 连同；以及；和……一起
pasture [ˈpæstʃər] *n.* 牧场；草原 *vi.* 吃草 *vt.* 放牧
operation [ˌɑːpəˈreɪʃn] *n.* 操作，经营
docile [ˈdɑːsl] *adj.* 温顺的；驯服的；易驾驭的；驯化
instinct [ˈɪnˌstɪŋkt] *n.* 本能，天性；冲动；*adj.* 深深地充满着
process [ˈproʊses] *n.* 过程 *v.* 处理
evaluation [ɪˌvæljʊˈeʃən] *n.* 评估，定值，估计
comparison [kəmˈpærɪsən] *n.* 比较，对照
contemporary [kənˈtempəreri] *adj.* 当代的；属一个时期的
ideally [aɪˈdiəli] *adv.* 理想地
accuracy [ˈækjərəsi] *n.* 精确（性），准确（性）
certified [ˈsɜːtəˌfaɪd] *adj.* 被鉴定的；被证明的；有保证的；公认的
consist of [kənˈsɪst ʌv] 包括；由……组成；由……组成

4. Notes to the Text A

(1) Seed stock producers should have well defined breeding goals and objectives for each breed or line of swine that they raise.

不同品系和品种的种猪具备不同的性能，比如长白猪和大白猪具有良好的繁殖性能，而杜洛克猪做父本最好。因此种畜养殖场应该“因材施教”，扬长避短地对种猪进行育种选择。

(2) Other producers rotate several breeds of dual purpose sires in the production of market hogs and replacement gilts.

有一种生产模式，养殖户在生产商品肉猪和后备母猪的过程中，交替使用用于这两种用途的父本来配种。

(3) A breeder selling animals to herds with sows on pasture might have a slightly different selection emphasis than a breeder who sells boars and gilts to confinement operations.

有一些养殖户是对母猪进行户外散养的，而另一些养殖户是对猪只进行室内圈养的。针对不同的养殖模式，需要制订出相对的适应该模式的育种计划，而不是一味地套上固定的育种计划。

(4) Records should be collected on most of the herd and processed in a timely manner by a genetic evaluation program so it is possible to make meaningful comparisons.

在饲养过程中，畜群的性能记录是很有必要的，我们需要收集这些数据，对数据采用遗传评估程序进行分析和比较，根据变化从而及时地调整育种计划，以保证能让计划最优化。

(5) Furthermore, a contemporary group should have no more than a three-to four-week span in ages to help reduce environmental differences.

环境差异对畜群的影响需要尽量控制，因此在选择同代畜群时应考虑其年龄跨度不能超过3~4周，否则猪只受到的环境效应影响过大，会导致育种计划出现偏差。

5. Exercises

(1) Fill in the blanks by finishing the sentences according the passage

①Seedstock producers should have well defined ________ and ________ for each breed or line of swine that they raise.

②Unique ________ ________ should also be considered.

③ In determining selection objectives, breeders might also consider the

________ used by potential customers.

④Seedstock producers need to produce ________ possible for their customers.

⑤ ________ are important for accurate genetic evaluations.

⑥ ________ and ________ that allow collection of accurate records should be used.

（2） Answer the following questions according to the passage

①What kind of breeding goals and objectives should be suppied by seedstock producers?

②What should be considered when we decide the type of crossbreeding system being used by potential customers?

③How do breeders reply the different production methods used by potential customers?

④Why do breeders need to organize an effective within herd selection program?

⑤What are across herd genetic ties?

⑥After reading this article, how do you think to make a well defined breeding goals and objectives?

（3） Translation of the following sentences into Chinese

①These goals should fit the strengths (or niche) of each breed and be designed to meet specific needs of targeted markets.

②Some commercial producers mate lean, fast growing boars (terminal sires) to prolific crossbred females (maternal lines) with all resulting pigs going to market.

③Opportunities exist to market pork that excels in meat quality.

④Total confinement with slatted floors, hoop structures with bedded floors, and outdoor farrowing are examples of some different production methods.

⑤Having a reasonable size contemporary group is important for reliable genetic evaluations.

⑥Producers should use testing methods which consists of recording all litters and performance testing at least 50% of the pigs weaned.

课文A 种畜生产者

种畜生产者应该为他们饲养的每个品种或品系提供明确的育种目标。这些目标应符合每个品种的优势，旨在满足目标市场的特定需求。各种杂交选育计划、市场机会和潜在客户不同的生产模式（室内和户外生产）等不确定因素，都需要使用不同的选择目标。

在决定强化哪些性状、哪些品种间可进行杂交时，应把潜在客户使用的杂交系统类型考虑进去。一些养殖户将瘦肉率高、生长速度快的公猪（终端父本）与产仔数高的杂交母猪（母系）进行配种，最后所有的猪只都流向市场。一些养殖户在生产商品肉猪和后备母猪的过程中交替使用双用途父本品种。用于生产终端父本的品种应该注重强化断乳后的性状。在母系和双用途品种中，生产者应该对繁殖和断乳后的性状进行组合选择。

还应考虑独特的市场机会。市场上对肉质好的猪肉有大量需求。对于肉质优良的品种，生产者的育种选择计划应该把肉质性状与其他重要性状一起考虑在内。

在确定选择目标时，育种者也应该考虑潜在客户使用的生产模式。使用板条地板制造的总分娩舍，使用层状地板制造的围栏结构以及户外分娩都是不同生产模式的例子。将母猪群卖给户外放养模式的猪场和将公猪和后备母猪卖给室内饲养模式的猪场需要制定不同的选择对象。对于户外作业的养殖客户而言，饲养员可能会将动物脾性视为选择计划中需要增加的一个额外性状。脾性的选择在这里变得十分重要，因为这些户外养殖的客户需要温顺的母猪，这种母猪需要具有良好的母性本能、最少的人工照料。这种饲养模式还需要活跃的公猪，用于牧场上的栏位配种。

种畜生产者需要为其客户提供最好的种畜。为实现这一目的，需要制订一个有效的选择计划。大多数选择计划同时包括猪群内选择和户外公猪及精液选择。必须妥善建立起一个有效的畜群内选择计划。应收集大多数畜群的记录，并通过遗传评估程序及时处理数据，以便对数据进行对比。在选择动物时，当育种者对同一代的畜群结构进行设计时就可能发现一些有意义的数据比较。一个设计合理的同代畜群里的动物具有相同的性别和生存环境。同代畜群至少应该有 20 头猪只，包括来自 5 窝的猪只和 2 头及以上的公猪。理想情况下，这些公猪中的一只能被其他育种者使用，从而实现跨群遗传连接。

跨群连接对于准确的遗传评估非常重要。通过购买精液，可以接触到相关公猪并为畜群之间提供遗传连接。具有一定规模的同代畜群对遗传评估的可靠性非常重要。此外，为减少环境差异，同代畜群的年龄跨度不应超过 4 周。

使用能收集到准确数据的设备和技术。通过利用经国家猪改良联合会认证的实时超声技术，性能测试的准确性得到了提高。生产者应该使用包括所有窝数和记录至少 50%断奶仔猪的性能测试的测试方法。记录的数据应由遗传评估程序及时处理。育种者应该通过这些记录数据来选择最好的种猪，以取代较低等级的公猪和母猪。最后，育种者还要考虑它们的配种问题，防止近亲交配。

Part Two Extensive Reading

1. Text B Genomic Selection Shows Promise in Swine Breeding

Genetic improvement has played a role in improving nearly every production efficiency trait evaluated in livestock, including pigs. The rapid evolution of gene technology allows swine breeders and commercial pork producers to make breeding decisions based on gene marker technology, once thought to be applicable only to researchers leading the "genome revolution".

Pig breeders have been using gene marker technology since the early 1990s to remove deleterious genes like the halothane gene (HAL), which causes porcine stress syndrome, and the napole gene (RN−) from their herds.

Several new gene marker tools are now commercially available at relatively low costs for traits including number of pigs born, feed efficiency, growth rate, backfat depth and pork quality. These tools are but the first wave of the genomic technology that will continue to revolutionize pork production efficiency.

Evolution of Quantitative Tools

For many years, pork producers relied on the physical attributes and growth rate measures (phenotypic evaluation) to evaluate the sires of their next pig crop. Later, quantitative genetics and selection indexes were employed to more accurately estimate an animal's genetic merit by using a number of economically important traits and a limited number of relative's records.

In the 1950s and 1960s, testing stations sprang up to standardize the environment in which boars' performance was measured and indexed. This information helped guide producers toward sires that would improve the productivity of their herds. Development of the backfat probe improved measurement accuracy and, hence, lean percentage.

Soon, whole-herd testing helped identify family lines that consistently excelled in economically important production traits. A new statistical procedure called best linear unbiased prediction (BLUP), used in mixed models for the prediction of random effects, allowed breeders to take the performance information from numerous individuals and combine it with the genetic relationship between the animals. This accounting for the relationships between animals was a major advancement in estimating genetic merit.

During the 1970s and 1980s, advancements in ultrasonic technology provided measurement of a variety of traits, such as backfat, loin muscle area and depth and intra-muscular fat to help predict lean percentage and meat quality traits.

Commercial producers can now conduct a genetic evaluation of their sow herd to identify females with the most desirable estimated breeding values (EBVs) to produce replacement gilts. Breeders commonly use more complex computer tools, such as mate selection programs, to optimize genetic progress and track inbreeding accumulations.

2. Notes to the Text B

Commercial producers can now conduct a genetic evaluation of their sow herd to identify females with the most desirable estimated breeding values (EBVs) to produce replacement gilts.

估计育种值（EBV）是现代育种对性状进行评估的一个重要指标。如今的商业生产中，通过EBV对母猪进行遗传评估已经成为必不可少的一个环节。

3. Answer Questions

(1) What's the Genomic Selection?

(2) Do you know the evolution process of quantitative tools?

(3) How to estimate an animal's genetic merit more accurately?

(4) What was a major advancement in estimating genetic merit?

(5) When is ultrasonic technology provided measurement of a variety of traits?

课文B　基因组选择在猪育种中的应用前景

遗传改良对提高牲畜（包括猪）的几乎所有生产效率性状都起到了重要作用。基因技术的迅速发展使得猪育种者和商业猪肉生产者能够基于基因标记技术做出育种决定，该技术曾被认为仅适用于领导“基因组革命”的研究人员。

自20世纪90年代早期以来，猪育种人员一直使用基因标记技术去除有害基因，例如导致猪应激综合征的氟烷基因（HAL）和来自其畜群的RN-基因。

现在有几种新的基因标记工具可以以相对较低的成本购得，可用于标记产仔数、饲料效率、生长速度、背膘厚度和猪肉质量。这些工具只是基因组技术应用的第一波，它将继续革新猪肉生产效率。

量化工具的演变

多年来，猪肉生产者依靠身体属性和生长速度测定（表型评估）来评估它们的后代。后来，通过使用一些重要的经济性状和有限数量的亲属记录，采用数量遗传学和选择指数来更准确地评估动物的遗传价值。

在20世纪50年代和60年代，测定站开始被使用，公猪的表型能在这个环境中进行标准测定，并加入到指数中。这些信息帮助指导生产者改善种群的生产力。背膘测定探头的开发提高了测量精度，从而提高了瘦肉率。

很快，全群测定帮助确定了在重要经济性状上长期表现优异的家族。一种称为最佳线性无偏预测（BLUP）的新统计程序，用于预测随机效应的混合模型，允许育种者从众多个体中获取性能信息，并将其与动物之间的遗传关系结合起来。对动物之间这种关联考虑是估计遗传价值的重大进步。

在20世纪70年代和80年代，超声波技术的进步使得多种性状可测量，如背膘，眼肌面积和深度以及肌内脂肪，这些有助于预测瘦肉率和肉质性状。

商业化生产者现在可以对其母猪进行遗传评估，以鉴定具有最理想估计育种值（EBV）的母猪以生产后备母猪。育种人员通常使用更复杂的计算机工具，如选配计划，以优化遗传进展并追踪近交累积。

扫码进行拓展学习

Lesson 3
Reproduction

Part One Intensive Reading

1. Learning Objective

After learning this lesson, you should understand the following:

(1) The most commonly used method to artificially inseminate cattle.

(2) The reason why you use your left hand in the rectum to manipulate the reproductive tract and the right hand to manipulate the insemination gun.

(3) How to avoid startling or surprising the animal?

(4) How to insert the gun into the vagina successfully?

(5) Where is the cervix located in most cows?

(6) To become a successful inseminator, it is very important that you always know where the tip of the insemination gun is located.

2. Text A Artificial Insemination

The recto-vaginal technique is the most commonly used method to artificially inseminate cattle. The basic skills required to perform this technique can be obtained with about three days practice under professional instruction and supervision. Additional proficiency and confidence will be achieved with further work on your own.

Regardless of whether you are left or right handed, it is recommended that you use your left hand in the rectum to manipulate the reproductive tract and the right hand to manipulate the insemination gun. The reason is that the rumen or stomach of the cow lies on the left side of the abdominal cavity, displacing the reproductive tract slightly to the right. Thus, you will find it much easier to locate and manipulate the tract with your left as opposed to right hand.

A gentle pat on the rump or a soft-spoken word as you approach for insemination, will help to avoid startling or surprising the animal. Raise the tail with your right hand and gently massage the rectum with the lubricated glove on your left hand. Place the tail on the back side of your left forearm so it will not interfere with the insemination process. Cup your fingers together in a pointed fashion and insert your hand in the rectum, up to the wrist.

Gently wipe the vulva with a paper towel to remove excess manure and debris. Be careful not to apply excessive pressure, which may smear or push manure into the vulva and vagina. With your left hand make a fist and press down directly on top of the vulva. This will spread the vulva lips allowing clear access to insert the gun tip several centimeter into the vagina before contacting the vaginal walls. Insert the gun at a 30° upward angle to avoid entering the urethral opening and bladder located on the floor of the vagina. With the gun about 15~20 cm inside the vagina, raise the rear of the gun to a somewhat level position and slide it forward until it contacts the external portion of the cervix. You will note a distinct gristly sensation on the gun when it contacts the end of the cervix.

The cervix, consists of dense connective tissue and muscle and is your primary landmark for inseminating cattle. It has often been described as having the size and consistency of a turkey neck. The size will vary, with post partum interval and age of the animal. The cervix usually has three or four annular rings or folds. The opening into the cervix protrudes back into the vagina. This forms a 360° blind-ended pocket completely around the cervical opening. This pocket is referred to as the fornix. In most cows, the cervix will be located on the floor of the pelvic cavity near the anterior end of the pelvic bone. In older cows with large reproductive tracts, the cervix may rest slightly over the pelvic bone and down into the abdominal cavity.

To become a successful inseminator, it is very important that you always know where the tip of the insemination gun is located. The walls of the vagina consist of thin-layered muscle and loose connective tissue. The insemination gun can be easily felt with your palpating hand.

3. New Words and Phrases

vagina [vəˈdʒaɪnə] *n.* 阴道

artificially inseminate 人工授精

instruction [ɪnˈstrʌkʃən] *n.* 指令，命令；指示；教导；用法说明

supervision [sʊpərˈvɪʒən] *n.* 监督；管理；监督的行为、过程或作用

rectum [rɛktəm] *n.* 直肠

rumen [ˈruːmen] *n.* 瘤胃（反刍动物的第一胃），胃液
vulva [vʌlvə] *n.* 阴户，女阴；孔
manure [məˈnʊr] *n.* 肥料 *vt.* 施肥
inseminating gun [ɪnˈsemɪneɪt] [gʌn] *n.* 输精枪
urethral [jʊˈriθrəl] *adj.* [解剖] 尿道的
bladder [blædər] *n.* 膀胱
sensation [senˈseɪʃn] *n.* 感觉；轰动；激动；知觉
wrist [rɪst] *n.* 手腕；腕关节 *vt.* 用腕力移动
cervix [ˈsɜːvɪks] *n.* 子宫颈；颈部

4. Notes to the Text A

(1) Regardless of whether you are left or right handed, it is recommended that you use your left hand in the rectum to manipulate the reproductive tract and the right hand to manipulate the insemination gun.

无论你是左撇子还是右撇子，我们都推荐你使用左手进入直肠把握生殖道，而用右手操作输精枪。

(2) Place the tail on the back side of your left forearm so it will not interfere with the insemination process. Cup your fingers together in a pointed fashion and insert your hand in the rectum, up to the wrist.

将牛尾放于左手外侧，避免在输精过程中影响你的操作。并拢左手手指形成锥形，缓缓进入直肠，直至手腕位置。

(3) Be careful not to apply excessive pressure, which may smear or push manure into the vulva and vagina. With your left hand make a fist and press down directly on top of the vulva. This will spread the vulva lips allowing clear access to insert the gun tip several inches into the vagina before contacting the vaginal walls.

在擦的过程中不要太用力，以免将粪便带入生殖道。左手握拳，在阴门上方垂直向下压。这样可将阴门打开，输精枪头在进入阴道时不要与外门壁接触，避免污染。

(4) It has often been described as having the size and consistency of a turkey neck. The size will vary, with post partum interval and age of the animal.

通常人们形容子宫颈的大小和硬度像火鸡的脖子。但对于不同年龄和产后不同时期的牛，其子宫颈的大小有所差异。

(5) In most cows, the cervix will be located on the floor of the pelvic cavity near the anterior end of the pelvic bone. In older cows with large reproductive tracts, the cervix may rest slightly over the pelvic bone and down into the abdominal

cavity.

对于绝大多数牛，子宫颈位于骨盆腔靠近盆骨前缘。但对于生殖道较粗的老年牛，子宫颈可能会轻度向前坠入腹腔。

(6) To become a successful inseminator, it is very important that you always know where the tip of the insemination gun is located.

要想成为一名成功的配种员，就必须自始至终明确输精枪头的位置，这一点很重要。

5. Exercises

(1) Fill in the blanks by finishing the sentences according to the passage

①The recto-vaginal technique is the most commonly used method to ________ ________ cattle.

②It is recommended that you use your left hand in the rectum to manipulate the ________ ________ and the right hand to manipulate the ________ .

③Raise the tail with your right hand and gently massage the rectum with the ________ on your left hand.

④Gently wipe the vulva with a paper towel to remove excess ________ and ________ .

⑤The cervix, consists of dense ________ tissue and muscle and is your ________ landmark for inseminating cattle.

⑥The cervix usually has three or four ________ rings or folds.

⑦The walls of the ________ consist of thin-layered muscle and loose connective tissue.

(2) Answer the following questions according to the passage

①Why is it recommended that you use your left hand in the rectum to manipulate the reproductive tract and the right hand to manipulate the insemination gun?

②What should you do when insert your hand in the rectum?

③When did the restraint practices of animals become important?

④Why not to apply excessive pressure when you use a paper towel to remove excess manure and debris?

⑤What does fornix mean?

⑥What are the important for a successful inseminator ?

(3) Translation of the following sentences into Chinese

①The basic skills required to perform this technique can be obtained with about three days practice under professional instruction and supervision.

②A gentle pat on the rump or a soft－spoken word as you approach for insemination, will help to avoid startling or surprising the animal.

③With the gun about 15~20 cm inside the vagina, raise the rear of the gun to a somewhat level position and slide it forward until it contacts the external portion of the cervix.

④In most cows, the cervix will be located on the floor of the pelvic cavity near the anterior end of the pelvic bone. In older cows with large reproductive tracts, the cervix may rest slightly over the pelvic bone and down into the abdominal cavity.

⑤The walls of the vagina consist of thin－layered muscle and loose connective tissue. The insemination gun can be easily felt with your palpating hand.

课文A 人工授精

直肠把握法是牛在人工授精中最普遍采用的一种方法。经过专业指导和培训，一般可在三天时间内基本掌握这一方法的操作要领，但熟练程度和自信心的提高则需要个人更多的实践。

无论你是左撇子还是右撇子，我们都推荐你使用左手进入直肠把握生殖道，而用右手操作输精枪。这是因为奶牛的瘤胃位于腹腔的左侧，将生殖道轻微地推向了右侧。所以你会发觉用左手要比用右手更容易找到和把握生殖道。

在你靠近牛准备人工授精的时候，轻拍牛的臀部或温和的呼唤牛将有助于避免牛受到惊吓。先将输精手套套在左手，并用润滑液润滑，然后用右手举起牛尾，左手缓缓按摩外门。将牛尾放于左手外侧，避免在输精过程中影响你的操作。并拢左手手指形成锥形，缓缓进入直肠，直至手腕位置。

使用一张纸巾擦去阴门外的粪便。在擦的过程中不要太用力，以免将粪便带入生殖道。左手握拳，在阴门上方垂直向下压。这样可将阴门打开，输精枪头在进入阴道时不要与外门壁接触，避免污染。斜向上30°插入输精枪，避免枪头进入位于阴道下方的输尿管口和膀胱内。当输精枪进入阴道15~20cm，将枪的后端适当抬起，然后向前推至子宫颈外口。当枪头到达子宫颈时，你能感觉到一种感觉完全不同的软组织顶住输精枪。

子宫颈由致密结缔组织和肌肉构成，是牛在人工授精过程中的重要节点。通常人们形容子宫颈的大小和硬度像火鸡的脖子。但对于不同年龄和产后不同时期的牛，其子宫颈的大小有所差异。子宫颈内通常有3~4个折叠环。子宫颈的开口向阴道突出，与阴道内壁形成一个360°闭合的穹窿结构，专业术语称之为阴道穹窿。对于绝大多数牛，子宫颈位于骨盆腔靠近盆骨前缘。但对于生殖道较粗的老年牛，子宫颈可能会轻度向前坠入腹腔。

要想成为一名成功的配种员，就必须自始至终明确输精枪头的位置，这一点很重要。阴道壁是由薄的肌肉层和疏松结缔组织构成的，所以你可以很方便地用手触摸到输精枪。当输精枪进入阴道后，你可以让枪与触诊的手平行前进。

Part Two Extensive Reading

1. Text B Future Challenges in Pig Reproduction

Reproductive efficiency remains central to the pig production industry, not only for the obvious function of perpetuating the species, but also to provide opportunities to reduce the costs of production while meeting increasingly demanding consumer expectations. In recent years the balance of pig reproduction research has shifted away from a primary objective to increase the number of piglets born per year, towards strategies that also promote piglet viability and which optimise the lifetime performance of the breeding sow. This trend is likely to continue and will require a more integrated approach in which the opportunities provided by increased understanding of reproductive processes are considered as an essential component of a co-ordinated strategy to improve pig production.

Compared to other domestic species, the refinement and uptake of assisted reproductive technologies in pigs has proved particularly challenging. Recent developments in semen sorting coupled with the ability to inseminate directly into the uterus using substantially reduced doses of semen mean that such technologies are becoming within reach of the industry. The considerable developments in embryo technologies in recent years, although partially prompted by biomedical applications, provide new opportunities for the long-term storage and global transport of pig genetic material. These technologies and their applications are particularly timely, given increasing concerns over biosecurity and the conservation of animal genetic resources.

Litter size, the interval between weaning and conception and mortality remain major areas where reproductive performance can be improved. While considerable research effort has been spent optimising mean values of these traits within specific production systems, relatively less has been devoted to reducing the inherent variability in these traits. For example, while increased litter size and/or mean birth weight have been major research goals, the within-litter variability in piglet weight at birth has received less attention, despite the fact that such variation poses considerable labour, welfare and management issues.

Future reproduction research should be directed towards the production of predictable numbers of healthy, uniform piglets able to cope with the challenges presented by contemporary production systems.

2. Notes to the Text B

(1) In recent years the balance of pig reproduction research has shifted away from a primary objective to increase the number of piglets born per year, towards strategies that also promote piglet viability and which optimise the lifetime performance of the breeding sow.

近年来，猪繁殖研究开始考虑平衡，由提高母猪年产仔猪数作为首要目标转向提高仔猪生存能力和优化母猪终生繁殖性能的策略。“from…toward”为“从……转向……”，“strategies”后面跟随由“that”和“which”引导的从句。

(2) While considerable research effort has been spent optimising mean values of these traits within specific production systems, relatively less has been devoted to reducing the inherent variability in these traits.

虽然已经花费了相当多的研究工作来优化特定生产系统中这些性状的平均值，但在减少这些性状内在变异方面投入的相对较少。

句中“Considerable”表示“相当多的”，“relatively less”为定语前置。

3. Answer Questions

(1) What remains central to the pig production industry?

(2) What has proved particularly challenging in pigs compared to other domestic species?

(3) What gives the concerns over biosecurity and the conservation of animal genetic resources?

(4) What are the major areas to improve the reproductive performance?

课文B 猪繁殖的未来挑战

繁殖效率仍然是养猪业的核心，这不仅是因为具有延续物种的明显功能，而且还可以提供一种机会，在降低生产成本的同时增加消费者的期望。近年来，猪繁殖研究开始考虑平衡，由提高母猪年产仔猪数作为首要目标转向提高仔猪生存能力和优化母猪终生繁殖性能的策略。这种趋势可能会持续下去，通过增加对生殖过程的了解采取更加综合的方法，将是改善养猪生产战略的重要组成部分。

与其他家畜相比，猪辅助生殖技术的改进和优化的工作尤其具有挑战性。精液分选的最新发展以及使用低精液剂量进行子宫内输精的技术将很快被行业所认可使用。近年来胚胎技术的长足发展，虽然部分是由生物医学应用推动的，但为猪遗传物质的长期储存和全球装运提供了新的机遇。这些技术及其应用尤其及时，因为人们对生物安全和动物遗传资源的保护日益关注。

产仔数、断奶与再配间隔和死亡率仍然是改善繁殖性能所重点关注的领域。虽然已经花费了相当多的研究工作来优化特定生产系统中这些性状的平均值，但在减少这些性状内在变异方面投入的相对较少。例如，虽然增加产仔数和/或平均出生重是主要的研究目标，但对出生时仔猪体重的变异关注较少，尽管仔猪初生变异度造成了相当大的劳动、福利和管理问题。

未来的繁殖研究应该针对猪只可预测的健康、仔猪均匀度，才能够应对当前生产系统所面临的挑战。

扫码进行拓展学习

Lesson 4
Biotechnology

Part One Intensive Reading

1. Learning Objective

After learning this lesson, you should understand the following:

(1) Explain when cloning experiments have been taking place.

(2) Explain why an animal clone is defined as an animal that originates from another animal, and both animals share identical chromo-somal DNA.

(3) Give an overview of why SCNT is now the primary method used in animal cloning.

(4) Define animal science cloning efficiency.

(5) Describe what some clones are born with.

2. Text A Animal Cloning

Although recent advances have opened bountiful opportunities and discussions on animal cloning, cloning experiments have been taking place for more than 100 years.

An animal clone is broadly defined as an animal that originates from another animal, and both animals share identical chromosomal DNA. Hans Dreisch created the first animal clones in the late 1800s. He created sea urchin clones by splitting a two - cell embryo and allowing both cells to independently develop into sea urchins. These embryo - splitting experiments continued into the 1900s, led by the Nobel Prize winning Hans Spemman's work on salamander embryos. The next major advance came in 1952 when Robert Briggs cloned a frog using a new technique, he used nuclear transfer to transplant the nucleus of a blastomere from a frog embryo into an enucleated egg.

Although Briggs showed embryonic nuclear transfer could produce clones, not many believed that adult somatic cells could be used as donors. However, in 1996 the largest breakthrough in animal cloning came in the form of a sheep named Dolly. Dolly became the first animal to be cloned using the nucleus of a differentiated adult cell as a donor. Dolly opened the door to cloning via somatic cell nuclear transfer (SCNT), and many other species have been cloned in the last few decades. SCNT is now the primary method used in animal cloning. The procedure first begins by extracting oocytes from a female donor and allowing the oocytes to mature in vitro. Once an oocyte has matured the nucleus can be removed using a needle aspiration system. The enucleated oocyte is now ready to accept the donor cell.

There remains healthy debate concerning whether the donor cell should undergo a serum starvation treatment before being inserted into the oocyte, as well as the significance of the type and age of the cell used. The serum starvation treatment arrests the donor cell in the G_0 phase, stopping further division. Once a donor cell is selected it can be inserted under the zona pellucida of the oocyte. The two cells are fused together by a brief electrical stimulus, which is referred to as electrofusion. The developmental and directing factors of the ooplasm reprogram the somatic nucleus to develop into an embryo and eventually a blastocyst, after which it can be transferred into a recipient.

The potential advantages of cloning are innumerable for many industries including agriculture and biomedical research. However, the field is still relatively new and needs extensive research to make animal cloning more efficient. Cloning efficiency is defined as the number of live offspring per embryos transferred. Currently, the efficiency rate for cattle is 6%~15% and 6% for pigs. However, for some animals the efficiency is as low as 1%~2%, whereas others still have not been successfully cloned. Although cloning efficiency has improved in the past 10 years, these proportions are still substantially lower than other reproductive techniques.

In addition, some clones are born with phenotypic abnormalities. The most common abnormality is an unusually increased birth weight, known as large offspring syndrome (LOS). LOS causes difficulties in the birthing process, as well as other health risks for the animal, such as organ defects and diabetes. These abnormal phenotypes are not transmitted to the clone's offspring, which suggest in vitro conditions alter the epigenetic patterns of the cloned embryo, as these patterns are reprogrammed during gametogenesis.

These in vitro conditions are being studied to help improve efficiency and reduce abnormalities in animal cloning, as a number of biological factors are known to influence the reprogramming of the nucleus.

3. New Words and Phrases

split [splɪt] *v.* 分裂，破裂；*n.* 裂缝，裂开，破裂；*adj.* 分裂的，不一致的

embryo [ˈembrioʊ] *n.* 胚胎，胎儿

independently [ɪndɪˈpɛndəntli] *adv.* 独立地

sea urchin [si] [ɜːrtʃɪn] 海胆

salamander [sæləmændər] *n.* 蝾螈

frog [frɔːg] *n.* 青蛙

transplant [trænsplænt] 移植

somatic cell 体细胞

donors [dəʊnə] *n.* 供体，施主，捐赠者，供血者，给予体

breakthrough [brekˈθrʊ] *n.* 发现，突破

somatic cell nuclear transfer （SCNT）体细胞核移植

extract [ˈekstrækt] *v.* 提取，抽出，开采，摘录

oocyte [ˈoursəɪt] *n.* 卵母细胞

debate [dɪˈbeɪt] *v.* 争辩，争论，辩论；*n.* 争论，争议，辩论会，议论

undergo [ʌndərˈgoʊ] *v.* 经历，经受，遭受，意会

serum [sɪrəm] *n.* 血清，浆液

starvation [stɑːrˈveɪʃn] *n.* 饥饿，饿死

significance [sɪgˈnɪfəkəns] *n.* 显著性

phase [feɪz] *n.* 时期，阶段，局面，段落，学时

division [dɪˈvɪʒn] *n.* 分裂，部门，除法，部类

zona pellucida 透明带

brief [briːf] *adj.* 短暂的，简短的，简要的，简略的；*n.* 简报，短简；*v.* 交代

refer [rɪˈfɜːr] *v.* 简称，参考，参照，谈到，提交

electrofusion [elektrɒfˈjuːʒn] *n.* 电熔，电融合

ooplasm [ˈoʊrplæzm] *n.* 卵胞质

eventually [ɪˈventʃuəli] *adv.* 终于，终究，竟

blastocyst [ˈblæstəsɪst] *n.* 胚泡，囊胚

innumerable [ɪˈnuːmərəbl] *adj.* 无数的，不计其数，不可胜数

biomedical [ˌbaɪoʊˈmedɪkl] *adj.* 生物医药

offspring [ɔfsprɪŋ] *n.* 子孙，后裔

phenotypic [ˌfiːnəʊˈtɪpɪk] *adj.* 表型的
abnormality [ˌæbnɔːrˈmæləti] *n.* 异常
large offspring syndrome LOS 胎儿过大综合征
organ [ɔrgən] *n.* 器官
diabetes [daɪəˈbitiz] *n.* 糖尿病
transmit [trænzˈmɪt] *v.* 传递，遗传，传导，播放，发送，发射
epigenetic [ˌepɪdʒɪˈnetɪk] *n.* 表观遗传学
gametogenesis [gəˌmitəˈdʒɛnɪsɪs] *n.* 配子

4. Notes to the Text A

（1） SCNT is now the primary method used in animal cloning. The procedure first begins by extracting oocytes from a female donor and allowing the oocytes to mature in vitro. Once an oocyte has matured, the nucleus can be removed using a needle aspiration system. The enucleated oocyte is now ready to accept the donor cell.

此处提出了体细胞核移植（SCNT）概念，并做出详细解释。

（2） The serum starvation treatment arrests the donor cell in the G_0 phase, stopping further division.

此处提出细胞分裂的 G_0 期，要熟悉细胞分裂各个时期的主要特点。

（3） In addition, some clones are born with phenotypic abnormalities. The most common abnormality is an unusually increased birth weight, known as large offspring syndrome（LOS）. LOS causes difficulties in the birthing process, as well as other health risks for the animal, such as organ defects and diabetes.

LOS 是克隆所带来的表型异常，会导致某些负面的影响。

（4） These abnormal phenotypes are not transmitted to the clone's offspring, which suggest in vitro conditions alter the epigenetic patterns of the cloned embryo, as these patterns are reprogrammed during gametogenesis.

此句中的“epigenetic”原意是后生，在此处是指表观遗传学。

5. Exercises

（1） Fill in the blanks by finishing the sentences according to the passage

①An animal clone is broadly defined as an animal that originates from another animal, and both animals share identical chromo-somal ________ .

②Hans Dreisch created the first animal clones in the late ________ .

③Dolly opened the door to cloning via ________ , and many other species have been cloned in the last few decades.

④________ is now the primary method used in animal cloning.

⑤The ________ treatment arrests the donor cell in the G_0 phase, stopping further division. Once a donor cell is selected it can be inserted under the zona pellucida of the oocyte. The two cells are fused together by a brief electrical stimulus, which is referred to as ________.

⑥________ is defined as the number of live offspring per embryos transferred.

(2) Answer the following questions according to the passage

①What is somatic cell nuclear transfer?

②What is the definition of electrofusion?

③What are the most common phenotypic abnormalities at birth of some cloned animals?

④What is the definition of cloning efficiency?

(3) Translation of the following sentences into Chinese

①Hans Dreisch created the first animal clones in the late 1800s. He created sea urchin clones by splitting a two-cell embryo and allowing both cells to independently develop into sea urchins.

②The procedure first begins by extracting oocytes from a female donor and allowing the oocytes to mature in vitro. Once an oocyte has matured the nucleus can be removed using a needle aspiration system. The enucleated oocyte is now ready to accept the donor cell.

③The developmental and directing factors of the ooplasm reprogram the somatic nucleus to develop into an embryo and eventually a blastocyst, after which it can be transferred into a recipient.

④LOS causes difficulties in the birthing process, as well as other health risks for the animal, such as organ defects and diabetes.

课文A 克隆

尽管关于动物克隆是在最近开放了很多的机会，并进行了热烈的讨论，但克隆试验实际已经有了100多年历史。

动物克隆的广义是指一个动物来源于另一个动物，这两个动物拥有完全相同的非同一般染色体的DNA。Hans Dreisch在19世纪末创造了第一个克隆动物。他通过两个胚胎细胞的分裂，并使这两个细胞各自独立生长成海胆，从而创造了海胆克隆。这些胚胎分裂试验在20世纪继续发展，使得Hans Spemman关于蝾螈目动物的胚胎细胞移植工作获得了诺贝尔奖。接下来主要的发展是在1952年，当时Robert Briggs用一项新的技术克隆了青蛙；他使用了从青蛙囊胚细胞中取出细胞核移植到一个去核卵细胞中的核移植技术。

尽管Briggs展示了可以通过胚胎细胞核移植技术来进行克隆，但是没有多少人相信成熟体细胞能作为供体。然而，在1996年关于动物克隆的一项最重大的突破是克隆羊“多莉”。“多莉”羊成为第一个使用分化成熟细胞的细胞核作为供体进行克隆的动物。“多莉”羊打开了通过体细胞核移植（SCNT）克隆的大门，并且在过去的几十年里，越来越多的其他物种通过这项技术被成功克隆。SCNT是目前动物克隆最主要的方法。这项技术首先是从雌性供体中提取卵母细胞，并使卵母细胞在体外成熟。一旦卵母细胞成熟，就可以用针吸取细胞核来移除细胞核。被去除细胞核的卵母细胞就可以接受供体细胞了。

关于供体细胞是否应在进入卵母细胞前进行血清饥饿，还有关于细胞的类型及供体年龄等方面的重要性仍存在健康方面的争议。血清饥饿使得供体细胞停滞在G_0期，停止进一步的分裂活动。一旦选择了供体细胞，它就能够附着到卵母细胞的透明带中。这两个细胞通过一个简短的电刺激融合在一起，这一过程被称为电融合。体细胞的细胞核在卵母细胞胞质内生长并重组，发育成为胚胎，最终形成囊胚，然后可将其移植入受体体内。

克隆在农业、生物医学研究等许多行业中都有无数潜在的优点。然而，这一领域目前仍然是相对比较新的，还需要进行广泛的研究来使得动物克隆更加高效。克隆效率被定义为每个胚胎移植后的后代活体数量。目前，牛的克隆效率在6%~15%，猪的克隆效率在6%。而有些动物的克隆效率低至1%~2%，甚至还有一些其他的动物还没能成功克隆。尽管克隆效率在过去的10年里有了提高，但是克隆动物的占比仍然大大低于其他的生殖技术。

另外，一些克隆动物出生时表型异常。最常见的是出生体重异常增加，称为胎儿过大综合征（LOS）。胎儿过大综合征（LOS）导致分娩过程比较困难，以及动物体的一些其他健康风险，例如器官缺陷和糖尿病。这些异常表型不会遗传给克隆体的后代，这表明是体外的条件改变了克隆胚胎的表观遗传模式，因为这些模式在配子发生期间被重新编辑了。

这些体外条件目前正在研究中，以帮助提高克隆效率，并减少动物克隆异常，目前已知有众多生物因素会影响细胞核的重编辑。

Part Two Extensive Reading

1. Text B Production of Transgenic Animals

Inserting human genes into an animal's genome allows animals to produce important human proteins, such as the blood clotting agent factor IX. However, the methods for producing transgenic animals are not very efficient. Incorporation rates of the new gene into their genome are low and occur at random sites which often do not allow the gene to be expressed. Also, the insertion can cause disruption in the expression of another gene. Researchers at the Roslin Institute sought to use cloning as an efficient way to produce transgenic animals. In theory, once a cell line successfully incorporates and expresses a transgene, that cell line can be used as a donor cell for cloning. The clones produced will have the transgene incorporated into their genome and can successfully pass it to their offspring through traditional breeding methods. This could lead to entire herds of transgenic animals expressing important genes for medical and agricultural purposes. Transgenics and cloning also hold enormous potential for producing organs in animals for human transplants, or xenotransplantation. If animals can be modified to produce viable organs for humans, cloning could drastically increase the human organ supply.

Many transgenic technologies are inefficient because they involve nonspecific integration of the transgene into the target genome. In contrast, nuclease-mediated genome editing results in a specific integration. The method relies on the use of artificial proteins made up of customizable, sequence-specific DNA-binding domains fused to a nuclease that cuts DNA in a nonsequence specific way. Zinc-finger nucleases (ZFNs) and recently, TALENS are employed in performing targeted genome editing. ZFNs and TALENS can be described as molecular scissors that cleave double—stranded DNA at a specific site in a predetermined sequence of the genome. The cleavage triggers DNA repair that can be exploited to modify the genome either by targeted introduction of insertions and deletions (gene disruption), base substitution specified by a homologous donor DNA construct (gene correction), or the transfer of entire transgenes into a native genomic locus. This new technique has the potential to be used in many applications including therapeutic approaches to treat genetic disease, production of model organisms, and generation of new agriculturally relevant varieties.

2. Notes to the Text B

(1) Many transgenic technologies are inefficient because they involve nonspecific integration of the transgene into the target genome. In contrast, nuclease-mediated genome editing results in a specific integration. The method relies on the use of artificial proteins made up of customizable, sequence-specific DNA-binding domains fused to a nuclease that cuts DNA in a nonsequence specific way.

此处详细介绍并解释了一种新的、更有效的转基因技术。

(2) ZFNs and TALENS can be described as molecular scissors that cleave double-stranded DNA at a specific site in a predetermined sequence of the genome.

此处将“ZFNs”和“TALENS”两种酶形象地比喻为“剪刀”，使用了比喻的修辞手法。

(3) This new technique has the potential to be used in many applications including therapeutic approaches to treat genetic disease, production of model organisms, and generation of new agriculturally relevant varieties.

此句对这种新技术给予充分肯定，也对其能带来的效果充满希望。

3. Answer Questions

(1) How can animal produce important human proteins?

(2) Which cell line can be used as a donor cell for cloning?

(3) Why are many transgenic technologies inefficient?

(4) What is the nuclease-mediated genome editing?

(5) What can ZFNs and TALENS be described as?

课文B 转基因动物的生产

将人类基因插入到动物基因组，允许动物产生重要的人类蛋白质，例如凝血剂因子Ⅸ。但是，这种使用转基因动物生产的方法不是很有效。新基因在它们的基因组中的并入率低，并且经常发生在一些不允许基因表达的随机位点上。另外，插入也可能导致其他基因的表达遭到破坏。Roslin研究所的研究人员试图将克隆作为一种有效的生产转基因动物的方法。理论上来说，一旦一个细胞系成功并入并表达转基因，该细胞系即可以被用作克隆的供体细胞。克隆产物可将转基因并入它们的基因组中，并能够通过传统的育种方法成功地遗传给后代。这可能导致全部转基因动物群体表达医疗和农业所需要的重要基因。转基因和克隆也都具有使用动物来生产用于人体移植或异种移植的器官的巨大潜力。如果动物能够被用来生产可用的人体器官，那么可大大提高人体器官的供应量。

许多转基因技术效率很低，因为它们涉及非特异性地整合转基因到目标基因组中。相反，核酸酶介导的基因编辑则会导致特定的整合。这种方法依赖于使用定制的人造蛋白，序列特异性 DNA 结合域以非序列特异性方式来切割 DNA 与核酸酶融合。锌指核酸酶（ZFNs）和最近的 TALENS 被用于执行靶向基因组编辑。ZFNs 和 TALENS 被描述为分子剪刀，因为它们可以在基因组的预定序列的特定位点切割双链 DNA。切割触发 DNA 的修复，其可以通过靶向插入或缺失（基因破坏），如基于同源供体 DNA 构建体指定的碱基置换（基因校正），或将整个转基因转移到天然的基因组位点中的方式来进行基因组的修饰。这种新技术有可能被应用于治疗遗传疾病、模型生物的生产、新农业相关产品的生产等多项领域中。

扫码进行拓展学习

Unit Ⅱ Nutrition and Feeding

Lesson 5 Nutrition

Part One Intensive Reading

1. Learning Objective

After learning this lesson, you should study and understand the following:

(1) Give an overview of the name carbohydrate is derived from.

(2) Explain the reason that the vitamins to classify to water-soluble vitamins and fat-soluble vitamins.

(3) Describe the functions of calcium and phosphorus in the body.

(4) Define the essential amino acid.

2. Text A The Components of Animal Feed

Carbohydrates

The name carbohydrate is derived from the fact that these compounds contain carbon, hydrogen, and oxygen. Hydrogen and oxygen are nearly always present in the ratio in which they are found in water (2 : 1, H_2O). Sugars are the first physiological products of photosynthesis and thus represent the basis for the only renewable source of food energy. The photosynthetic reaction may be represented as follows:

$$6\ CO_2 + 6\ H_2O + 2816kJ = C_6H_{12}O_6 + 6\ O_2$$

The carbohydrates are usually classified as monosaccharides, disaccharides, and polysaccharides, based on the multiples of five- or six-carbon sugars contained in the molecule. The hexoses (six-carbon sugars) and pentoses (five-carbon sugars) comprise the monosaccharides. The most important hexoses are glucose, fructose, galactose, and mannose. Most disaccharides are sugars with the general formula $C_{12}H_{22}O_{11}$, which upon hydrolysis yield 2 moles of hexose. This group of the carbohydrate family represents higher molecular weight complex combinations of lesser sugars.

Lipids

The lipid group (fats, ether-extractible) contains naturally occurring substances characterized by their insolubility in water and their solubility in such fat solvents as ether, chloroform, boiling alcohol, and benzene. This group includes not only the true fats but also materials that are related chemically (lecithin) and materials that have comparable solubility properties (cholesterol and waxes). The true fats are of interest not only because of their concentrated source of energy (2.25 times that of carbohydrates and proteins) but also because a number of vitamins are associated with fat. These are the fat-soluble vitamins (vitamin A, vitamin D, vitamin E, and vitamin K). In addition, most nutritionists recognize the need for certain so-called essential fatty acids (linoleic and linolenic).

Fats and oils are esters of fatty acids with glycerol; waxes are esters of fatty acids with alcohols other than glycerol. Upon hydrolysis, a typical fat yields three molecules of fatty acid and one molecule of glycerol. Hydrolysis of fat in the presence of alkali (sodium or potassium) results in the formation of alkali salts of the respective fatty acids or soaps (saponification).

Proteins

Proteins are defined as complex nitrogen-containing organic compounds found in all animal and vegetable cells. On hydrolysis proteins yield approximately 23 amino acids. Characteristically, proteins exist as large molecules or perhaps clusters of molecules. Corn protein (zein) has a molecular weight of over 35,000 compared to the carbohydrate glucose, which has a molecular weight of 180. Hemoglobin has a molecular weight greater than 70,000. Hydrolysis of proteins yields a number of intermediate products, namely, proteoses and peptones.

Proteins differ in the number, kind, and arrangement of amino acids in the molecule. The amino acids obtained from protein are typically of the alpha form. The amino group ($—NH_2$) is attached to the carbon atom adjacent to the carboxyl group ($—COOH$), with two exceptions, proline and hydroxyproline. Although many more than 23 amino acids have been shown to occur in nature, and many more have been synthesized, only the 23 are accepted as building blocks in protein synthesis.

An essential amino acid is defined as an amino acid that an animal cannot synthesize in sufficient quantities for normal functioning. Theoretically, the ruminating animal can synthesize all of its amino acid requirements if it is supplied with all of the necessary elements. On the other hand, there are approximately 10 amino acids that cannot be synthesized in sufficient quantities. These 10 amino acids are thus "essential" for the monogastric animal (11 in the case of poultry): arginine, histidine, isoleucine, leucine, lysine, methionine (plus cystine), phenylalanine (plus tyrosine), threonine, tryptophan, and valine. In the case of poultry, the amino acid glycine (plus serine) must be added to this list. (The amino acids listed in parentheses can supply a portion of the amino acid with which they are associated in the list.)

Minerals

The mineral group is defined as that portion of the diet that will not be destroyed by ignition under extremely high temperature conditions. It is this technique that is employed in assaying materials for ash content. Any such nonspecific assay has obvious built-in errors. For example, extremely high temperatures will drive off much of the volatile fluorine. The cooling of certain minerals, such as iron, encourages oxidation, and thus the weight of such ash could be increased slightly due to at least minimal addition of the oxygen of oxidation.

Most scientists are of the opinion that some 15 mineral elements are functional. However, many more are found in the body, and many are considered to have functions that are not yet understood.

Calcium: About 99% of the calcium of the body is found in the skeletal structure; thus, less than 1% is found in soft tissues. However, this small fraction of the total calcium found in soft tissues is of tremendous importance. For example, excitability of the muscles, rhythm of the heart muscle, clotting of blood, maintenance of acid-base equilibrium, regulation of the permeability of membranes, and activation of certain enzymes all are dependent upon calcium.

Phosphorus: Phosphorus is found in all body cells and fluids and occurs in nearly every feedstuff. Metabolism of phosphorus is so interrelated with that of calcium that many authors discuss the two together. Phosphorus, in combination with calcium, forms the supportive skeleton of vertebrates. It is concerned with fat metabolism through its involvement in the formation of lecithin and also plays a key role in carbohydrate's metabolism through the formation of hexose phosphates, adenylic acid, and creatine phosphate. Phosphates play an integral part in the absorption of sugars from the small intestine and reabsorption of glucose from the kidney tubules. Phosphorus is also found in nervous tissues.

About 80% of body phosphorus is found in the skeleton, 10% in the muscles, and 1% in the nervous system. The remainder is generally distributed throughout body cells. Within the cell, most of the phosphorus is found in the nucleus. Red blood cells and plasma are rich in phosphorus. Vitamins B are phosphoric acid compounds.

Vitamins

The vitamins actually have very little in common, as far as chemical structure is concerned: some are "sugar" acids, some are sterols, some contain nitrogen, some contain no nitrogen, some are fat-soluble, and some are water-soluble. It is on the basis of the solubility characteristics that the vitamin group is divided, namely, those that are fat-soluble and those that are water-soluble. Here, much of the similarity within each group ends.

Vitamins may be defined as organic substances required by animals for normal life and functioning. They cannot be synthesized in sufficient quantities, and are similar to a catalytic agent in that they are effective in very small quantities but are not sources of energy and are not utilized in the structure of an animal. Vitamins have a great deal in common with hormones in the manner in which they operate, but are different in that the hormones originate from the endocrine system of the body.

Very little is understood concerning the physiological function of any of the vitamins. Therefore, their functions are defined as preventives for certain dysfunctions; for example, vitamin C is called the antiscurvy vitamin.

3. New Words and Phrases

carbohydrate [kɑːbəˈhaɪdreɪt] *n.* 碳水化合物；糖类

physiological [ˌfɪzɪəˈlɒdʒɪkəl] *adj.* 生理学的，生理的

photosynthesis [ˌfəʊtəʊˈsɪnθɪsɪs] *n.* 光合作用

renewable [rɪˈnjuːəbl] *adj.* 可再生的；可更新的；可继续的；*n.* 再生性能源

reaction [rɪˈækʃən] *n.* 反应，感应；反动，复古；反作用

monosaccharide [mɒnəˈsækəraɪd] *n.* 单糖，单糖类（最简单的糖类）

disaccharide [daɪˈsækəraɪd] *n.* 二糖
polysaccharide [ˌpɒlɪˈsækəraɪd] *n.* 多糖；多聚糖
hexose [ˈheksəʊz] *n.* 已糖
pentose [ˈpentəʊz] *n.* 戊糖
glucose [ˈgluːkəʊs] *n.* 葡萄糖
fructose [ˈfrʌktəʊz] *n.* 果糖；左旋糖
galactose [gəˈlæktəʊz] *n.* 半乳糖
mannose [ˈmænəʊz] *n.* 甘露糖
hydrolysis [haɪˈdrɒlɪsɪs] *n.* 水解作用
lipid [ˈlɪpɪd] *n.* 脂质；油脂
fat [fæt] *adj.* 肥的，胖的；油腻的；丰满的；*n.* 脂肪，肥肉；*vt.* 养肥；在……中加入脂肪；*vi.* 长肥
solubility [ˌsɒljʊˈbɪlətɪ] *n.* 溶解度；可解决性
solvent [ˈsɒlvənt] *adj.* 有偿付能力的；有溶解力的；*n.* 溶剂；解决方法
ether [ˈiːθə] *n.* 乙醚；以太；苍天；天空醚
chloroform [ˈklɔːrəfɔːm] *n.* 氯仿；三氯甲烷；*vt.* 用氯仿麻醉
alcohol [ˈælkəhɔːl] *n.* 酒精，乙醇
benzene [ˈbenziːn] *n.* 苯
lecithin [ˈlesɪθɪn] *n.* 卵磷脂；蛋黄素
cholesterol [kəˈlestərɒl] *n.* 胆固醇
wax [waks] *n.* 蜡；蜡状物；*vt.* 给……上蜡；*vi.* 月亮渐满；增大；*adj.* 蜡制的；似蜡的
vitamin [ˈvɪtəmɪn] *n.* 维生素
essential fatty acid 必需脂肪酸
glycerol [ˈglɪsərɒl] *n.* 甘油；丙三醇
ester [ˈestə] *n.* 酯
alkali [ˈælkəlaɪ] *n.* 碱；可溶性无机盐；*adj.* 碱性的
sodium [ˈsəʊdɪəm] *n.* 钠
potassium [pəˈtæsɪəm] *n.* 钾
saponification [səˌpɒnɪfɪˈkeɪʃən] *n.* 皂化
protein [ˈprəʊtiːn] *n.* 蛋白质；朊；*adj.* 蛋白质的
nitrogen-containing organic compound 含氮有机化合物
amino acid 氨基酸
molecular weight 相对分子质量
hemoglobin 血红蛋白

intermediate product 中间产物

proteose [ˈprəʊtɪəʊs] *n.* 朊间质；（蛋白）脲

peptone [ˈpeptəʊn] *n.* 蛋白胨，胨

proline [ˈprəʊliːn] *n.* 脯氨酸

hydroxyproline [haiˌdrɔksiˈprəuliːn] *n.* 羟（基）脯氨酸

synthesis [ˈsɪnθəsɪs] *n.* 综合，合成；综合体

element [ˈelɪmənt] *n.* 元素；要素；原理；成分；自然环境

monogastric animal 单胃动物

arginine [ˈɑrdʒəˌnin] *n.* 精氨酸

histidine [ˈhɪstɪdiːn] *n.* 组氨酸

isoleucine [ˌaɪsəʊˈluːsiːn] *n.* 异亮氨酸

leucine [ˈluːsiːn] *n.* 亮氨酸；白氨酸

lysine [ˈlaɪsiːn] *n.* 赖氨酸

methionine [mɪˈθaɪəniːn] *n.* 蛋氨酸

cystine [ˈsɪstiːn] *n.* 胱氨酸；双硫丙氨酸

phenylalanine [ˌfiːnaɪlˈælәniːn] *n.* 苯基丙氨酸

tyrosine [ˈtaɪrəsiːn] *n.* 酪氨酸

threonine [ˈθriːəniːn] *n.* 苏氨酸；羟丁胺酸

tryptophan [ˈtrɪptəfæn] *n.* 色氨酸

valine [ˈveɪliːn] *n.* 缬氨酸

glycine [ˈglaɪsiːn] *n.* 甘氨酸；氨基乙酸

serine [ˈsɪəriːn] *n.* 丝氨酸

mineral [ˈmɪnərəl] *n.* 矿物；（英）矿泉水；无机物；苏打水（常用复数表示）；*adj.* 矿物的；矿质的

ignition [ɪgˈnɪʃən] *n.* 点火，点燃；着火，燃烧；点火开关，点火装置

ash [æʃ] *n.* 粗灰分；灰；灰烬

volatile [ˈvɒlətaɪl] *adj.* 挥发性的；不稳定的；爆炸性的；反复无常的；*n.* 挥发物；有翅的动物

fluorine [ˈflʊəriːn] *n.* 氟

iron [ˈaɪən] *n.* 熨斗；烙铁；坚强；*vt.* 熨；用铁铸成；*adj.* 铁的；残酷的；刚强的；*vi.* 熨衣；烫平

oxidation [ɒksɪˈdeɪʃən] *n.* 氧化

oxygen [ˈɒksɪdʒən] *n.* 氧气，氧

calcium [ˈkælsiəm] *n.* 钙

skeletal [ˈskelətl] *adj.* 骨骼的，像骨骼的；骸骨的；骨瘦如柴的

soft tissue 软组织
excitability 兴奋性
rhythm [ˈrɪðəm] *n.* 节奏；韵律
clotting [ˈklɔtiŋ] *n.* 凝血；结块；*v.* 结块（clot 的 ing 形式）
acid-base equilibrium 酸碱平衡
permeability [pɜːmɪəˈbɪlɪtɪ] *n.* 渗透性；透磁率，导磁系数；弥漫
membrane [ˈmembreɪn] *n.* 膜；薄膜；羊皮纸
activation [ˌæktɪˈveɪʃn] *n.* 激活；活化作用
enzyme [ˈenzaɪm] *n.* 酶
phosphorus [ˈfɒsfərəs] *n.* 磷
feedstuff [ˈfiːdstʌf] *n.* 饲料（等于 feedingstuff）
metabolism [məˈtæbəlɪzəm] *n.* 新陈代谢
vertebrate [ˈvɜːtɪˌbreɪt] *adj.* 脊椎动物的；有脊椎的；*n.* 脊椎动物
hexose phosphates 磷酸己糖
adenylic acid 腺苷酸
creatine phosphate 磷酸肌酸
phosphate [ˈfɒsfeɪt] *n.* 磷酸盐；皮膜化成
absorption [əbˈzɔːrpʃn，əbˈsɔːrpʃn] *n.* 吸收；全神贯注，专心致志
small intestine 小肠
kidney tubules 肾小管
nervous tissue 神经组织
nucleus [ˈnjuklɪəs] *n.* 细胞核
red blood cell 红细胞
plasma [ˈplæzmə] *n.* 等离子体；血浆
catalytic agent 催化剂
endocrinesystem 内分泌系统
dysfunction [dɪsˈʃʌŋkʃən] *n.*（关系、行为等的）不正常，异常；功能障碍
antiscurvy [ˌæntiˈskɜːrvi] *n.* 抗坏血病

4. Notes to the TextA

(1) The true fats are of interest not only because of their concentrated source of energy (2. 25 times that of carbohydrates and proteins) but also because a number of vitamins are associated with fat.

句中“not only…, but also …”为连接词，意为“不仅……，而且……”。

(2) Corn protein (zein) has a molecular weight of over 35, 000 compared to

the carbohydrate glucose, which has a molecular weight of 180.

句中 “compared to” 表比较。 “which” 引导非限制性定语从句, 修饰 “carbohydrate glucose”。

(3) Theoretically, the ruminating animal can synthesize all of its amino acid requirements if it is supplied with all of the necessary elements.

句中 “if” 引导条件状语从句。

(4) About 80% of body phosphorus is found in the skeleton, 10% in the muscles, and 1% in the nervous system.

此句为省略句, 10%和1%后省略了 “of body phosphorus is found”。

(5) They cannot be synthesized in sufficient quantities, and are similar to a catalytic agent in that they are effective in very small quantities but are not sources of energy and are not utilized in the structure of an animal.

句中 “and” 引导并列句。“but” 表转折。

5. Exercises

(1) Fill in the blanks by finishing the sentences according to the passage

①The name carbohydrate is derived from the fact that these compounds contain ________, ________, and ________.

②The true fats are of interest not only because of their concentrated source of energy (2. 25 times that of ________ and ________) but also because a number of ________ are associated with fat.

③Upon hydrolysis, a typical fat yields three molecules of ________, and one molecule of ________.

④On hydrolysis proteins yield approximately 23 ________.

⑤Corn protein (zein) has a molecular weight of over 35, 000 compared to the ________, which has a molecular weight of 180.

⑥Proteins differ in the ________, and ________ of amino acids in the molecule.

⑦An essential amino acid is defined as an amino acid that an animal ________ synthesize in ________ quantities for normal functioning.

⑧The mineral group is defined as that portion of the diet that will not be destroyed by ________ under extremely high temperature conditions.

⑨About 99% of the calcium of the body is found in the skeletal structure; thus, less than 1% is found in ________.

⑩The vitamins actually have very little in common, as far as chemical structure is concerned: some are "sugar" acids, some are sterols, some contain nitrogen, some contain no nitrogen, some are ________, and some are water soluble.

(2) True or false

①In carbohydrate, hydrogen and oxygen are nearly always present in the ratio of 2 : 1. ( )

②The lipid group includes not only the true fats but also materials that are related chemically and materials that have comparable solubility properties. ( )

③The true proteins are of interest not only because of their concentrated source of energy (2.25 times that of carbohydrates and fats) but also because a number of vitamins are associated with fat. ( )

④Proteins are defined as complex oxygen-containing organic compounds found in all animal and vegetable cells. ( )

⑤The molecular weight of corn protein was more than of carbohydrate glucose. ( )

⑥The amino acids obtained from protein are typically of the alpha form. ( )

⑦More than 23 amino acids have been shown to occur in nature, but only the 23 are accepted as building blocks in protein synthesis. ( )

⑧An essential amino acid is defined as an amino acid that an animal can synthesize in sufficient quantities for normal functioning. ( )

⑨The small fraction of the total calcium found in soft tissues is of tremendous importance. ( )

⑩About 80% of body phosphorus is found in the skeleton, 10% in the nervous system, and 1% in the muscles. ( )

(3) Answer the following questions according to the passage

①How to represent the photosynthetic reaction?

②How to classify the carbohydrates?

③Which vitamins are included in the fat-soluble vitamins?

④What is the function of the small fraction of the total calcium found in soft tissues?

⑤What is the definition of vitamins?

(4) Translation of the following sentences into Chinese

① The lipid group (fats, ether-extractible) contains naturally occurring substances characterized by their insolubility in water and their solubility in such fat solvents as ether, chloroform, boiling alcohol, and benzene.

②Hydrolysis of fat in the presence of alkali (sodium or potassium) results in the formation of alkali salts of the respective fatty acids or soaps (saponification).

③The amino group (—NH_2) is attached to the carbon atom adjacent to the carboxyl group (—COOH), with two exceptions, proline and hydroxyproline.

④Although many more than 23 amino acids have been shown to occur in nature, and many more have been synthesized, only the 23 are accepted as building blocks in protein synthesis.

⑤The cooling of certain minerals, such as iron, encourages oxidation, and thus the weight of such ash could be increased slightly due to at least minimal addition of the oxygen of oxidation.

课文A　饲料成分

碳水化合物

碳水化合物包含碳、氢和氧元素。碳水化合物中氢和氧的比值与水中氢和氧的比值几乎相同（2∶1，H_2O）。糖是光合作用的第一个生理产物，因此是唯一可再生的食物能量来源。光合反应可表示为：

$$6\ CO_2 + 6\ H_2O + 2816kJ = C_6H_{12}O_6 + 6\ O_2$$

碳水化合物通常分为单糖、双糖和多糖，其结构单元为五碳糖或六碳糖。己糖（六碳糖）和戊糖（五碳糖）组成单糖。最重要的己糖是葡萄糖、果糖、半乳糖和甘露糖。大多数二糖的分子式为 $C_{12}H_{22}O_{11}$，水解后产生2分子的己糖。这类碳水化合物代表了较高分子量的复杂低糖组合。

脂肪

脂类（醚提取物）含有天然产生的物质，其特点是不溶于水，溶于乙醚、氯仿、沸腾的酒精和苯等脂类溶剂。这类物质不仅包括真正的脂肪，而且还包括与化学相关的物质（卵磷脂）和具有可溶性的物质（胆固醇和蜡）。真脂肪受关注不仅是因为它们是浓缩的能量来源（是碳水化合物和蛋白质的2.25倍），还因为许多维生素与脂肪有关。这些维生素是脂溶性维生素（维生素A，维生素D，维生素E和维生素K）。此外，许多营养学家认识到机体对必需脂肪酸（亚油酸和亚麻酸）的需要。

油脂是脂肪酸和甘油结合形成的酯类，蜡是脂肪酸和醇类结合形成的酯类。在水解过程中，典型的脂肪会产生3分子的脂肪酸和1分子的甘油。在碱（氢氧化钠或氢氧化钾）存在的条件下，脂肪水解会产生相应的脂肪酸或皂化的碱盐。

蛋白质

蛋白质被定义为在所有动物和植物细胞中发现的复杂含氮有机化合物。蛋白质水解大约产生23种氨基酸。典型蛋白质以大分子或分子簇的形式存在。玉米蛋白的相对分子质量为35000，而碳水化合物葡萄糖的相对分子质量为180。血红蛋白的相对分子质量大于70000。蛋白质水解产生许多中间产物，即蛋白脲和蛋白胨。

蛋白质在相对分子质量、种类和氨基酸排列上都有不同。从蛋白质中获得的氨基酸是典型的α排列方式。即氨基（$—NH_2$）与羧基（—COOH）附近的碳原子相连，但脯氨酸和羟脯氨酸例外。尽管在自然环境中发现了超过23种氨基酸，而且还有更多的氨基酸被合成，但只有23种氨基酸被认为是蛋白质合成中的构架。

必需氨基酸被定义为动物在正常功能情况下不能足量合成的氨基酸。理论上，如果提供了所有必需元素，反刍动物可以合成所有必需氨基酸。另一方面，大约有10种氨基酸不能被充足合成。因此，这10种氨基酸（在家禽中是11种）对于单胃动物来说是“必不可少的”：精氨酸、组氨酸、异亮氨酸、亮氨酸、赖氨酸、甲硫氨酸（加上胱氨酸）、苯丙氨酸（加上酪氨酸）、苏氨酸、色氨酸和缬氨酸。在家禽中，还必须添加甘氨酸（加上丝氨酸）。(括号中的氨基酸可以转化为一部分与列表相关的氨基酸。)

矿物质

矿物质被定义为在极高温条件下饲料中灼烧也不会被破坏的一部分物质。正是采用这种技术分析材料中的粗灰分。任何这样的非特异性试验都有明显的内在误差。例如，极高的温度会导致大部分挥发性氟元素消失。某些矿物质（如铁）的冷却会促进其氧化，因此，由于氧化过程中氧气的增加，这种粗灰分的质量会轻微增加。

大多数科学家认为大约有15种功能性矿物元素。然而，在机体中发现了更多矿物元素，其中许多矿物元素被认为具有的功能尚未确定。

钙（Ca)：机体中约99%的钙存在于骨骼结构中，而在软组织中有不到1%的钙。然而，在软组织中发现的这一小部分钙是非常重要的。例如，肌肉兴奋性、心肌节律、血液凝固、酸碱平衡的维持、细胞膜通透性的调节以及某些酶的激活都依赖于这部分钙。

磷（P)：在所有的机体细胞和体液中都能发现磷，几乎所有的饲料中都有磷。磷的代谢与钙的代谢密切相关，许多作者将两者结合在一起讨论。磷与钙结合形成脊椎动物的支撑性骨骼。它通过参与卵磷脂的形成而与脂肪代谢有关，并通过形成磷酸己糖、腺苷酸和磷酸肌酸对碳水化合物代谢起着关键作用。磷酸盐在小肠吸收糖和肾小管重吸收葡萄糖中起重要作用。磷也存在于神经组织中。

机体中80%的磷存在于骨骼中，10%的磷在肌肉中，1%的磷在神经系统中，其余的机体磷通常分布在体细胞中。在细胞内，大部分的磷都存在于细胞核中。红细胞和血浆富含磷。B族维生素是磷酸化合物。

维生素

维生素实际上没有什么共同之处，就化学结构而言：有些维生素是“糖”酸，有些维生素是固醇，有些维生素含氮，有些维生素不含氮，有些维生素是脂溶性的，有些维生素是水溶性的。根据维生素的溶解特征，分为脂溶性维生素和水溶性维生素。这样分类后，每类维生素大部分无相似性。

维生素是动物维持正常生活和生理功能所需要的有机物质。它们不能被足量合成，并且类似于一种催化剂，因为它们在数量很小时有效，但不是能量来源，也没有被动物结构利用。维生素与激素在作用方式上有很大的相似之处，但不同的是，激素来源于机体的内分泌系统。

目前人们对维生素的生理功能都了解较少。因此，它们的功能被定义为某些功能障碍的预防措施。例如，维生素C被称为抗坏血病维生素。

Part Two Extensive Reading

1. Text B Nutrients Requirements

Energy requirements

There are many sources of energy in the diet. Carbohydrates are the main source of energy. Fat is also a source of energy and research is underway to determine the value of adding fat to horse diets. Excess protein is utilized as energy but is an expensive source of energy and is not recommended for energy use. This unit will discuss all sources of energy and the horse's requirement for energy.

Owners of horses are very much concerned with the energy needs of their animals and are sometimes confused by the energy terms used in various reviews. The following scheme on energy terms may be helpful in understanding this subject.

Energy Breakdown during Digestion

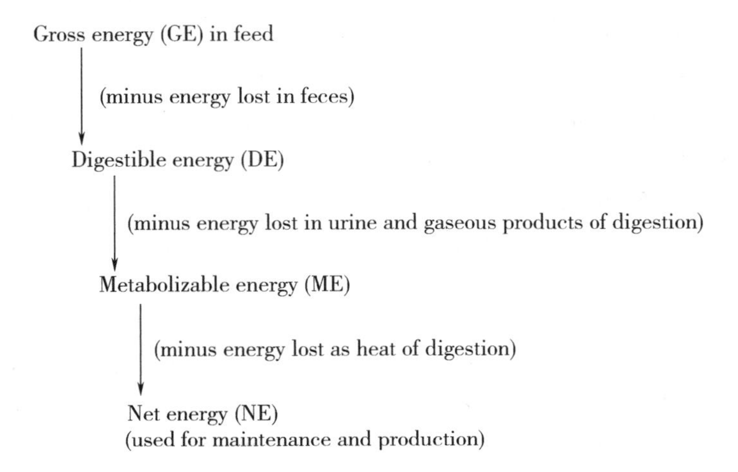

Brief explanations on the various kinds of energy are discussed below.

Gross energy: The amount of energy a feed will yield is measured in joules. It is usually reported as kilojoules (kJ) or mega joules (MJ). One kilojoule contains 1000 joules, and 1 MJ contains 1000 kJ. Fat has 39.56 kJ/g whereas protein has 23.65 kJ/g and carbohydrates 17.37 kJ/g.

Digestible energy: This is the gross energy minus the fecal energy (which is the energy not digested by the animal in the intestinal tract). The fecal energy is the greatest energy loss that occurs in the animal. In the horse, the loss will amount to 35%~40% of the gross energy in the feed.

Metabolizable energy: If the energy that is lost in the gaseous products of digestion and in the urine is subtracted from the digestible energy, it gives the metabolizable energy. In general, urine energy losses will amount to about 2%~5% of the gross energy in the feed for the horse. Very little experimental information is available on metabolizable energy levels in feeds for horses.

Net energy: There is another energy loss that occurs before metabolizable energy is available for productive purposes by the animal. This loss is called the heat increment by the scientist. The heat increment is the increase in heat production that occurs from the process of digestion and the breakdown of nutrients in the body. When the heat increment is subtracted from the metabolizable energy the remainder is net energy. Some of the heat increment can be used to keep the body warm. The net energy is used by the animal for maintenance, growth, reproduction, and lactation. Very little information is available on net energy values in feeds for horses.

A calorie (4.186 J) is defined as the amount of heat required to raise 1g of water 1 ℃ from 14.5 ℃ to 15.5 ℃. A kilocalorie (4.186 kJ) is the amount of heat required to raise 1000 g (i.e. 1 kg) of water 1 ℃ from 14.5 ℃ to 15.5 ℃.

Total digestible nutrients (TDN): Most horsemen are familiar with the term total digestible nutrients, but it is being used less now in scientific investigations. TDN in feeds equals the percentage of digestible protein plus percentage of digestible nitrogen-free extract plus percentage of digestible fat times 2.25.

One kilogram of TDN contains approximately 18418.4 kJ of digestible energy. Therefore, if one knows the TDN value of a feed, an estimated digestible energy value can be calculated from it. The 1989 NRC report states that energy values, such as TDN, DE, or NE, for feeds obtained from experiments with cattle may be significantly higher than those obtained from experiments with horses. Therefore, they state that such energy values should not be used directly in the formulation of horse diets, particularly if the feed contains a significant amount of fiber.

2. Notes to the Text B

(1) Fat is also a source of energy and research is underway to determine the value of adding fat to horse diets.

句中“be underway to”意为“正在进行”。

(2) The amount of energy a feed will yield is measured in joules.

句中“is measured”为被动语态。专业英语常用被动语态，原因是被动语态的结构比主动语态有更少的主观色彩，更可突出要说明的对象。

(3) A calorie is defined as the amount of heat required to raise 1 g of water 1℃ (1 degree centigrade) from 14.5℃ to 15.5℃.

句中“is defined”为被动语态。“of…”，“from…to”为介词短语，常用在专业英语中。

3. Answer Questions

(1) What are the sources of energy in the diet?

(2) How to break down energy during digestion?

(3) How to define the terms of gross energy, digestible energy, metabolizable energy and net enwrgy?

(4) How to define a calorie?

(5) What is total digestible nutrients (TDN)?

(6) How to calculate an estimated digestible energy value from TDN?

课文B　营养需要

能量需要

日粮中能量的来源很多。碳水化合物是主要的能量来源。脂肪也是能量的来源，现在正在进行研究，确定马日粮中添加脂肪的价值。过剩的蛋白质被用作能量，但它是一种昂贵的能源，不推荐被当作能量使用。本单元将讨论所有能量来源和马对能量的需要。

马的主人非常关心马的能量需要，有时会被多种评论中所使用的能量术语所迷惑。下面关于能量术语的描述有助于理解这一主题。

消化过程中能量的分解

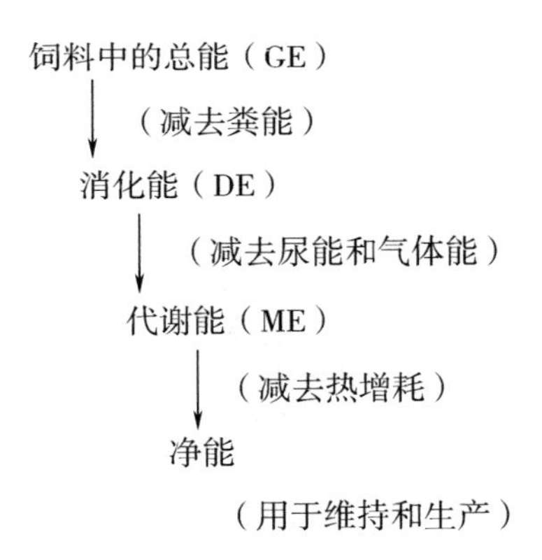

下面对各种能量进行简要的解释。

总能量：饲料产生的能量总和以焦耳来计算。一般以千焦（kJ）和焦耳（J）来表示。1千焦包含1000焦耳。每克脂肪有39.56kJ，每克蛋白质有23.65kJ，每克碳水化合物有17.37kJ。

消化能：消化能等于总能减去粪能（粪能是动物在消化道内为消化的能量）。粪能是动物最大的能量损失。对于马，粪能损失占饲料总能的35%~40%。

代谢能：代谢能等于消化能减去消化过程产生的气体能和尿能。一般情况下，尿能损失占马饲料总能的2%~5%。关于马饲料中代谢能水平的研究报道很少。

净能：还有一种能量损失发生在代谢能之前，被动物用于生产，科学家称之为热增耗。热增耗是机体营养物质的消化和分解而产生的热量的增加。代谢能减去热增耗就是净能。一部分热增耗被用于保持动物身体温暖。净能被用于动物维持、生长、繁殖、泌乳。关于马饲料中净能水平的研究报道非常少。

将1g水从14.5℃升高到15.5℃，即升高1℃所需要的热量定义为1卡。将1kg水从14.5℃升高到15.5℃，即升高1℃所需要的热量定义为1千卡。

总可消化养分：大多数养马人对总可消化养分这个词很熟悉，但这个词现在在科学研究中的使用越来越少。饲料中的总可消化养分等于可消化蛋白的百分比加可消化无氮浸出物的百分比再加上可消化脂肪的百分比乘以2.25。

1kg可消化养分包含约184184kJ消化能。因此，如果知道饲料的可消化养分值，就可以估算饲料中的消化能值。1989年NRC报道指出从牛的试验中获得的饲料能值如总可消化养分、消化能、净能可能显著高于从马的试验中获得的这些能值。因此，NRC指出从牛的试验中获得的这些能值不能直接用于配制马的日粮，尤其当饲料中包含大量纤维时。

扫码进行拓展学习

Lesson 6
Metabolism

Part One　Intensive Reading

1. Learning Objective

After learning this lesson, you should study and understand the following:

(1) Give an overview of why the most important step in carbohydrates metabolism is the oxidation to CO_2 and H_2O, with a concomitant release of energy.

(2) Describe the three fates of absorbed amino acids.

(3) Explain why fat is a readily available source of concentrated energy and can thus meet the demand.

2. Text A　Nutrient Metabolism

Carbohydrates Metabolism

Absorbed carbohydrates move rapidly to the liver, where they are stored, at least temporarily. The carbohydrates thus stored in the liver cells is deposited in the form of glycogen, commonly called animal starch. During such times, liver glycogen storage may equal 10% of the weight of the liver. However, this level may fall to practically zero when carbohydrates are not being absorbed. The muscles also store glycogen, and their content may reach 2% of their weight.

The immediate fate of glucose is regulated by the level of at least two hormones—insulin and epinephrine, also called adrenaline. Most researchers believe that insulin accelerates both conversion of glucose to glycogen and carbohydrate oxidation in the liver and muscles, and that adrenaline increases the rate of hydrolysis of liver glycogen to glucose and also the conversion of muscle glycogen to hexose phosphate.

The most important step in carbohydrate metabolism is the oxidation to CO_2 and H_2O, with a concomitant release of energy. The following equation is an oversimplified form of the net reaction:

$$C_6H_{12}O_6 + 6\ O_2 \longrightarrow 6\ CO_2 + 6\ H_2O + 2816kJ$$

Glucose is catabolized to pyruvate, which, in turn, enters the tricarboxylic cycle. By a series of reactions in which as many as a dozen enzymes participate, a molecule of glucose yields two molecules of three-carbon pyruvate and, eventually, two molecules of lactic acid. These final steps occur in the muscle under anaerobic conditions.

Of the three major fatty acids formed in the rumen, glucose is formed primarily from propionic acid. This transformation occurs in the liver. This reaction is critical to ruminant animals, which may consume only limited quantities of starch as a source of blood glucose. Butyric acid is also metabolized primarily in the liver, whereas acetic acid may bypass hepatic action and enter the peripheral circulation to be oxidized as an energy source. Acetic acid can be used by the mammary gland in producing milk fat. The three fatty acids may also be used as direct energy sources via the carboxylic acid cycle.

Lipid metabolism

When energy is needed quickly, especially when sufficient carbohydrates are not available, a condition called ketosis may develop in animals. This condition is quite common in pregnant ewes 1 or 2 weeks prior to lambing, especially in those carrying two or more fetuses. It can also occur in heavy milking cows shortly after parturition. Ketosis can be explained by the fact that the animal's body does not have sufficient carbohydrates to meet its energy demand. Thus, it breaks down body fat to meet the increased call for energy. Fat is a readily available source of concentrated energy and can thus meet the demand.

The acetone bodies (acetone, ketobutyric acid, β-hydroxybutyric acid, and acetoacetic acid) are normal products formed in the oxidation of fatty acids in the liver. However, the liver cannot utilize acetone bodies, and they are transported to the muscle for complete oxidation. Thus, if an overwhelming demand for energy requires excessive use of fat as the energy source, then more acetone bodies are formed, which are absorbed into the bloodstream. The bloodstream then contains an excessive amount of acetone bodies, the condition of ketosis is the result. Since the lungs can aerate a small quantity of the acetone bodies from the blood, the sickeningly sweet aroma of acetone on the breath of such animals becomes evident.

Protein and amino acids metabolism

The amino acids absorbed into the bloodstream are primarily required for the manufacture of tissue protein and other nitrogen-containing constituents, such as enzymes, milk, and eggs. Formation of tissue protein from amino acids is the reverse of the hydrolysis that is observed in the digestion process.

A second fate of absorbed amino acids is deamination. The kidneys and the liver both are capable of this process. The enzyme amino acid oxidase is involved in deamination. This enzyme is further subdivided into D-amino acid oxidases and L-amino acid oxidases, which act on the D-amino acids and L-amino acids, respectively. The latter form (L-amino acids) is found in feedstuffs. The amino acid is first dehydrogenated to the corresponding keto acid. The keto acid thus formed can be a converted to fat, b converted to carbohydrates, c resynthesized into an amino acid, and d oxidized to carbon dioxide and water.

A third fate of digested amino acids is storage. Although this capacity is quite limited compared to that of storage of fat, some capability for protein storage does exist. For example, heavy exercise can result in increased muscular bulk, which represents protein storage. Protein can also be stored briefly in blood.

3. New Words and Phrases

glycogen [ˈglaɪkədʒən] *n.* 糖原；动物淀粉

insulin [ˈɪnsjʊlɪn] *n.* 胰岛素

epinephrine [ˌepɪˈnefrɪn] *n.* 肾上腺素

pyruvate [paɪˈruːveɪt] *n.* 丙酮酸盐；丙酮酸酯

tricarboxylic cycle 三羧酸循环

lactic acid 乳酸

anaerobic [ˌænəˈroʊbɪk] *adj.* 厌氧的，厌气的；没有气而能生活的

rumen [ˈruːmen] *n.* 瘤胃（反刍动物的第一胃）

propionic acid 丙酸

ruminant animal 反刍动物

starch [stɑːtʃ] *n.* 淀粉；刻板，生硬；*vt.* 给……上浆

butyric acid 丁酸

acetic acid 醋酸

peripheral [pəˈrɪfərəl] *adj.* 外围的；次要的；（神经）末梢区域的；*n.* 外部设备

circulation [sɜːkjʊˈleɪʃən] *n.* 流通，传播；循环；发行量

mammary gland 乳腺

ketosis [kɪˈtəʊsɪs] *n.* 酮病

lambing [ˈlæmɪŋ] *adj.* 产羔羊的；*n.* 产羔羊
fetus [ˈfiːtəs] *n.* 胎儿，胎
parturition [ˌpɑːtjʊˈrɪʃən] *n.* 分娩，生产
acetone body 酮体
ketobutyric acid 酮基丁酸
β-hydroxybutyric acid β-羟丁酸
acetoacetic acid 乙酰乙酸
absorb [əbˈzɔːrbˌəbˈsɔːrb] *vt.* 吸收；吸引；承受；理解；使……全神贯注
bloodstream [ˈblʌdstriːm] *n.* 血流，血液的流动
deamination [di：ˌæmiˈneiʃən] *n.* 去氨基；脱氨基作用
kidney [ˈkɪdnɪ] *n.* 肾脏；腰子；个性
liver [ˈlɪvə] *n.* 肝脏；生活者，居民
dehydrogenate [diːhaɪˈdrɒdʒəneɪt] *vt.* 使脱氢
keto acid 酮酸
carbon dioxide 二氧化碳
digest [daɪˈdʒest] *vt.* 消化；吸收；融会贯通；*vi.* 消化；*n.* 文摘；摘要

4. Notes to the Text A

(1) Most researchers believe that insulin accelerates both conversion of glucose to glycogen and carbohydrate oxidation in the liver and muscles, and that adrenaline increases the rate of hydrolysis of liver glycogen to glucose and also the conversion of muscle glycogen to hexose phosphate.

句中“and”连接两个“that”引导的宾语从句。

(2) However, ingestion of protein has a more pronounced effect than ingestion of carbohydrates or fats.

句中“effect”指代上一句中的“elevates the metabolic rate”。

5. Exercises

(1) Fill in the blanks by finishing the sentences according to the passage

①The muscles also store ________, and their content may reach 2% of their weight.

②The most important step in carbohydrates metabolism is the oxidation to CO_2 and ________, with a concomitant release of ________.

③Of the three major fatty acids formed in the rumen, glucose is formed primarily from ________, .

④When energy is needed quickly, especially when sufficient carbohydrates are not available, a condition called ________ may develop in animals.

⑤This enzyme is further subdivided into ________ and L-amino acid oxidases, which act on the D-amino acids and L-amino acids, respectively.

(2) True or false

①The muscles also store glycogen, and their content may reach 2% of their weight. ( )

②Of the three major fatty acids formed in the rumen, glucose is formed primarily from lactic acid. ( )

③Acetic acid can be used by the mammary gland in producing milk fat. ( )

④When energy is needed slowly, especially when sufficient carbohydrates are not available, a condition called ketosis may develop in animals. ( )

⑤Ketosis can be explained by the fact that the animal's body have sufficient carbohydrate to meet its energy demand. ( )

⑥Fat is a readily unavailable source of concentrated energy and can thus meet the demand. ( )

⑦The lungs can aerate a large quantity of the acetone bodies from the blood. ( )

⑧The enzyme amino acid oxidase is not involved in deamination. ( )

(3) Answer the following questions according to the passage

①What is the most important step in carbohydrate metabolism?

②How to use the propionic acid, butyric acid, and acetic acid in metabolism?

③What are the three fates of absorbed amino acids?

④Which materials are the keto acid formed?

(4) Translation of the following sentences into Chinese

①Most researchers believe that insulin accelerates both conversion of glucose to glycogen and carbohydrates oxidation in the liver and muscles, and that adrenaline increases the rate of hydrolysis of liver glycogen to glucose and also the conversion of muscle glycogen to hexose phosphate.

②The amino acids absorbed into the bloodstream are primarily required for the manufacture of tissue protein and other nitrogen-containing constituents, such as enzymes, milk, and eggs.

③The keto acid thus formed can be a converted to fat, b converted to carbohydrates, c resynthesized into an amino acid, and d oxidized to carbon dioxide and water.

④It has been demonstrated that the addition of any of several amino acids to the diet, lysine in particular, increases the level of calcium absorption.

课文A　营养素代谢

碳水化合物代谢

吸收的碳水化合物会迅速转移到肝脏，并在肝脏储存起来，至少是暂时储存。储存在肝细胞中的碳水化合物以糖原的形式存在，通常被称为动物淀粉。在这种情况下，储备的肝糖原可能相当于肝脏质量的10%。然而，当碳水化合物未被吸收时，储备的肝糖原水平可能下降到几乎为零。肌肉也储存糖原，其含量可能达到肌肉质量的2%。

葡萄糖的代谢取决于至少两种激素，胰岛素和肾上腺素。大多数研究者认为，胰岛素加速了葡萄糖转化为糖原和碳水化合物在肝脏和肌肉中的氧化，而肾上腺素提高了肝糖原分解为葡萄糖的水解率，也使肌糖原转化为磷酸己糖。

碳水化合物代谢中最重要的步骤是将碳水化合物氧化，形成CO_2和H_2O，同时释放能量。下面的方程是碳水化合物代谢净反应的一个简化形式：

$$C_6H_{12}O_6 + 6\ O_2 \longrightarrow 6\ CO_2 + 6\ H_2O + 2817.08J$$

葡萄糖被分解成丙酮酸，进入三羧酸循环。在这一系列的反应中，有多达12种酶参与，一分子的葡萄糖产生两分子的三碳丙酮酸，最终，形成两分子的乳酸。这些步骤发生在处于厌氧条件下的肌肉。

在瘤胃中形成的三种主要脂肪酸中，葡萄糖主要由丙酸形成，这种转化过程发生在肝脏。这个反应对反刍动物至关重要，因为它们可能只消耗少量的淀粉作为血糖来源。丁酸也主要在肝脏中代谢，而乙酸则可以绕过肝脏的作用，进入外周血循环，作为能量来源被氧化。乙酸可被乳腺利用来分泌乳脂。这三种脂肪酸也可以通过羧酸循环来作为直接的能量来源。

脂肪代谢

当快速需要能量时，特别是当没有充足的碳水化合物时，动物会出现酮病。这种情况在产前1周或2周的怀孕母羊中非常常见，尤其是那些携带两个或更多胎儿的怀孕母羊。在分娩后不久的挤乳奶牛中也经常发生。酮病可以解释为，动物机体没有足够的碳水化合物来满足它的能量需要。因此，它分解机体脂肪以满足更多的能量需要。脂肪是一种容易获得的浓缩的能量来源，因此可以满足动物需要。

酮体（丙酮、酮基丁酸、β-羟丁酸和乙酰乙酸）是脂肪酸在肝脏氧化的正常产物。然而，肝脏不能利用酮体，它们被运送到肌肉中完全氧化。因此，如果对能量需要巨大，需要过度利用脂肪作为能量来源，就会形成更多的酮体，吸收到血液中。血液中含有过量的酮体，会导致酮病。由于肺能从血液中吸收少量的酮体，因此在这种动物的呼吸中会产生令人厌恶的芳香气味。

蛋白质和氨基酸代谢

血液中吸收的氨基酸主要用于构建组织蛋白和其他含氮组分，如酶、乳汁和卵细胞。氨基酸形成组织蛋白是在消化过程中观察到的蛋白质水解的逆反应。

被吸收的氨基酸的第二个作用是脱氨基。肾脏和肝脏都能完成这个过程。氨基酸氧化酶参与脱氨基。这种酶被进一步细分为D-氨基酸氧化酶和L-氨基酸氧化酶，分别作用于D-氨基酸和L-氨基酸。后一种形式（L-氨基酸）存在于饲料中。氨基酸首先脱氢形成相应的酮酸。由此形成的酮酸能被a转化为脂肪，b转化为碳水化合物，c再合成为一种氨基酸，d氧化形成二氧化碳和水。

被消化的氨基酸的第三个作用是储存。虽然与脂肪储存相比，氨基酸的储存能力相当有限，但确实可以储存一些蛋白质。例如，剧烈运动可以增加肌肉体积，代表蛋白质的储存。蛋白质也可以在血液中短暂储存。

Part Two Extensive Reading

1. Text B Trace Element Dynamics

Trace element dynamics in animals refers to the quantitative metabolism and kinetics of trace element absorption, distribution, storage and excretion. Control of these processes normally yields a reasonably constant and optimum internal environment with respect to trace element functions in metabolism. Within a range of conditions, trace element homeostasis may be achieved and maintained. Trace element dynamics may be viewed as the shifting or maintenance of trace element status, which depends on numerous factors including the element and species in question, and the influence of homeostatic mechanisms. It is the objective of this unit to summarize current knowledge of trace element dynamics with respect to the responses of whole-body trace element metabolism to changes in dietary intake.

Trace elements can be divided into two groups with respect to their route of endogenous excretion: those for which homeostasis is partially dependent on endogenous faecal excretion, controlled by the intestinal tract, liver and pancreas, and those for which homeostasis is dependent upon renal excretion, controlled by resorption in the proximal renal tubule. Cationic elements including Cu, Mn and Zn fall into the first category while elements present in the body as anions including Cr, F, Mo and Se fall into the second category. In the cation group, control of homeostasis through variation in absorption usually is the most significant factor, although variation in excretion is also important. Iron, though, is unique because its homeostasis is dependent essentially, if not entirely, on control of absorption. In the anionic group, control of excretion predominates and absorption plays a minor role. In this unit, we will discuss Zn, Cu, Mn, Fe and Se.

The trace element content of tissues may respond in basically two ways to changes in a dietary trace element intake: a no change over a range of intakes, beyond which a decrease or increase of tissue content occurs; or b a continuous change in tissue content over a range of intakes. In case a, a homeostatic plateau occurs, which is not apparent in case b. There are various combinations of tissue responses within individuals because some tissues maintain homeostasis more readily than others. Furthermore, there is a large variation among species. Responses to changes in dietary intake usually vary with level of intake, and several elements have specific storage tissues while others do not. One objective of this unit is to identify the variations in homeostatic responses of trace elements among tissues and species.

2. Notes to the Text B

(1) Trace element dynamics may be viewed as the shifting or maintenance of trace element status, which depends on numerous factors including the element and species in question, and the influence of homeostatic mechanisms.

句中“which”引导非限制性定语从句。“depends on”意为“取决于”。

(2) Trace elements can be divided into two groups with respect to their route of endogenous excretion: those for which homeostasis is partially dependent on endogenous faecal excretion, controlled by the intestinal tract, liver and pancreas, and those for which homeostasis is dependent upon renal excretion, controlled by resorption in the proximal renal tubule.

句中“those”指代“their route of endogenous excretion”。“and”引导并列

句。“controlled by…”为后置定语。

（3） Iron, though, is unique because its homeostasis is dependent essentially, if not entirely, on control of absorption. In the anionic group, control of excretion predominates and absorption plays a minor role.

句中“though”表转折。“if not”连词，引导否定的条件状语从句。

3. Answer Questions

（1） May trace element homeostasis be achieved and maintained within a range of conditions?

（2） Which groups of trace elements can be divided according to their route of endogenous excretion?

（3） Please state the ways of trace element content of tissues may respond to changes in a dietary trace element intake.

课文 B　微量元素动力学

动物的微量元素动力学是指微量元素吸收、分布、储存和排泄的定量代谢动力学。控制这些过程通常会产生关于新陈代谢中微量元素功能的恒定的和最佳的内部环境。在一定条件下，可以实现和维持微量元素内稳态。微量元素动力学可被看作是微量元素的转移或维持，它取决于许多因素，包括讨论中的元素和物种，以及稳态机制的影响。本单元的目的是总结目前关于整个机体微量元素代谢随着日粮摄入量变化的微量元素动力学的知识。

根据微量元素内源性排泄途径的不同可分为两类：第一类是稳态部分依赖于内源性粪便排泄，通过肠道、肝脏、胰腺调节；第二类是稳态依赖于肾排泄，由近端肾小管重吸收调节。阳离子元素包括铜、锰、锌，属于第一类。而在机体中存在的阴离子元素如铬、氟、钼、硒，属于第二类。在阳离子一类中，通过吸收的变化控制稳态通常是最重要的因素，虽然排泄的变化也很重要。铁由于对其稳态主要依赖于对吸收的控制而更为独特，虽然不是完全依赖。在阴离子一类中，对排泄的控制占主导地位，吸收作用较小。在本单元中，我们将讨论锌、铜、锰、铁和硒。

组织中的微量元素含量基本上可以通过两种方式对日粮微量元素摄入量的变化做出反应：a. 摄入量范围没变化，超出该范围，组织中的微量元素含量减少或增加；b. 超出摄入量范围，组织中的微量元素含量连续变化。如果是 a，会出现稳态平台期，这在 b 组中并不明显。在个体内存在多种组织反应的组合，因为一些组织比其他组织更容易保持稳态。此外，物种之间差异也很大。对日粮摄入量变化的反应通常随摄入量而变化，并且有些元素有特定的储存组织，而另一些则没有。本单元的一个目标是确定组织和物种中微量元素稳态反应的变化。

扫码进行拓展学习

Lesson 7
Animal Feed

Part One Intensive Reading

1. Learning Objective

After learning this lesson, you should study and understand the following:

(1) Give an overview of the functions of water in the body.

(2) Explain the fraction of feeds and their definition.

(3) Describe how to determine the moisture, ash, crude protein, crude fibre, ether extract, nitrogen-free extractives of food.

2. Text A The Animal and Its Food

Water

The water content of the animal body varies with age. The newborn animal contains 750 ~ 800 g/kg water but this falls to about 500 g/kg in the mature fat animal. It is vital to the life of the organism that the water content of the body be maintained: an animal will die more rapidly if deprived of water than if deprived of food. Water functions in the body as a solvent in which nutrients are transported about the body and in which waste products are excreted. Many of the chemical reactions brought about by enzymes take place in solution and involve hydrolysis. Because of the high specific heat of water, large changes in heat production can take place within the animal with very little alteration in body temperature. Water also has a high latent heat of evaporation, and its evaporation from the lung and skin gives it a further role in the regulation of body temperature.

The animal obtains its water from three sources: drinking water, water present in its food, and metabolic water, this last being formed during metabolism by the oxidation of hydrogen-containing organic nutrients. The water content of foods is variable and can range from as little as 60 g/kg in concentrates to over 900 g/kg in some root crops. Because of this great variation in water content, the composition of foods is often expressed on a dry matter basis, which allows a more valid comparison of nutrient content.

The water content of growing plants is related to the stage of growth, being greater in younger plants than in older plants. In temperate climates the acquisition of drinking water is not usually a problem and animals are provided with a continuous supply. There is no evidence that under normal conditions an excess of drinking water is harmful, and animals normally drink what they require.

Dry matter and its components

The dry matter (DM) of foods is conveniently divided into organic and inorganic material, although in living organisms there is no such sharp distinction. Many organic compounds contain mineral elements as structural components. Proteins, for example, contain sulphur, and many lipids and carbohydrates contain phosphorus.

The main component of the DM of pasture grass is carbohydrate, and this is true of all plants and many seeds. The oilseeds, such as groundnuts, are exceptional in containing large amounts of protein and lipid material. In contrast, the carbohydrate content of the animal body is very low. One of the main reasons for the difference between plants and animals is that, whereas the cell walls of plants consist of carbohydrate material, mainly cellulose, the walls of animal cells are composed almost entirely of lipid and protein. Furthermore, plants store energy largely in the form of carbohydrates such as starch and fructans, whereas an animal's main energy store is in the form of lipid.

The lipid content of the animal body is variable and is related to age, the older animal containing a much greater proportion than the young animal. The lipid content of living plants is relatively low, that of pasture grass, for example, being 40 ~ 50 g/kg DM.

In both plants and animals, proteins are the major nitrogen – containing compounds. In plants, in which most of the protein is present as enzymes, the concentration is high in the young growing plant and falls as the plant matures. In animals, muscle, skin, hair, feathers, wool and nails consist mainly of protein.

Like proteins, nucleic acids are also nitrogen–containing compounds and they play a basic role in the synthesis of proteins in all living organisms. They also carry the genetic information of the living cell.

The organic acids that occur in plants and animals include citric, malic, fumaric, succinic and pyruvic acids. Although these are normally present in small quantities, they nevertheless play an important role as intermediates in the general metabolism of the cell. Other organic acids occur as fermentation products in the rumen, or in silage, and these include acetic, propionic, butyric and lactic acids.

Vitamins are present in plants and animals in minute amounts, and many of them are important as components of enzyme systems. An important difference between plants and animals is that, whereas the former can synthesise all the vitamins they require for metabolism, animals cannot, or have very limited powers of synthesis, and are dependent upon an external supply.

The inorganic matter contains all those elements present in plants and animals other than carbon, hydrogen, oxygen and nitrogen. Calcium and phosphorus are the major inorganic components of animals, whereas potassium and silicon are the main inorganic elements in plants.

3. New Words and Phrases

mature [məˈtʃʊə] *adj.* 成熟的；充分考虑的；到期的；成年人的；*vi.* 成熟；到期；*vt.* 使……成熟；使……长成；慎重做出

waste product 排泄物

feed [fiːd] *vt.* 喂养；供给；放牧；抚养（家庭等）；靠……为生；*vi.* 吃东西；流入；*n.* 饲料；饲养；（动物或婴儿的）一餐

concentrates [ˈkɑnsɛnˌtret] *n.* 精矿；浓缩液（concentrate 的复数）；精料；浓缩物

dry matter 干物质

sulphur [ˈsʌlfə] *n.* 硫黄；硫黄色；*vt.* 使硫化；用硫黄处理

oilseed [ˈɒɪlsiːd] *n.* 含油种子（如花生仁、棉籽等）

groundnut [ˈgraʊndnʌt] *n.* 落花生；野豆

fructan [ˈfrʌktən] *n.* 果聚糖

body temperature 体温

nail [neɪl] *vt.* 钉；使固定；揭露；*n.* 指甲；钉子

nucleic acid 核酸

nitrogen-containing compound 含氮化合物

genetic information 遗传信息

organic acid 有机酸

citric acid 柠檬酸

malic acid 苹果酸

fumaric acid 延胡索酸

succinic acid 琥珀酸

pyruvic acid 丙酮酸

intermediate [ˌɪntəˈmiːdɪət] *n.* 媒介

fermentation product 发酵产物

minute [ˈmɪnɪt] *n.* 分，分钟；片刻，一会儿；备忘录，笔记；会议记录；*vt.* 将……记录下来；*adj.* 微小的，详细的

external supply 外源供应

component [kəmˈpəʊnənt] *adj.* 组成的，构成的；*n.* 成分；组件；[电子]元件

silicon [ˈsɪlɪkən] *n.* 硅；硅元素

proximate analysis 概略分析

moisture [ˈmɔɪstʃə] *n.* 水分；湿度；潮湿；降雨量

crude protein 粗蛋白

ether extract 粗脂肪

crude fibre 粗纤维

nitrogen-free extractive 无氮浸出物

silage [ˈsaɪlɪdʒ] *vt.* 青贮；把…… 放入青贮窖；*n.* 青贮饲料

4. Notes to the Text A

（1）The newborn animal contains 750~800 g/kg water but this falls to about 500 g/kg in the mature fat animal.

句中“but”表转折，“this”指代“water”。

（2）It is vital to the life of the organism that the water content of the body be maintained: an animal will die more rapidly if deprived of water than if deprived of food.

句中“it is… that”为强调句。

（3）Water functions in the body as a solvent in which nutrients are transported about the body and in which waste products are excreted.

句中“which”指代“solvent”，“and”引导并列句。

（4）The water content of foods is variable and can range from as little as 60 g/kg in concentrates to over 900 g/kg in some root crops.

省略句，“over 900 g/kg in some root crops”前省略了“the water content of foods”。

（5）The lipid content of living plants is relatively low, that of pasture grass, for example, being 40~50 g/kg DM.

句中“that”指代前半句中的“the lipid content”。

(6) In plants, in which most of the protein is present as enzymes, the concentration is high in the young growing plant and falls as the plant matures.

句中“and”引导并列句。“falls as the plant matures”前省略“the concentration”。

(7) Although these are normally present in small quantities, they nevertheless play an important role as intermediates in the general metabolism of the cell.

句中“nevertheless”表转折。

(8) An important difference between plants and animals is that, whereas the former can synthesise all the vitamins they require for metabolism, animals cannot, or have very limited powers of synthesis, and are dependent upon an external supply.

句中“the former”指代“plants”。“animals cannot”后省略了“synthesise all the vitamins they require for metabolism”。

5. Exercises

(1) Fill in the blanks by finishing the sentences according to the passage

①The newborn animal contains 750~800 g/kg ________ but this ________ to about 500 g/kg in the ________ fat animal.

②Water functions in the body as a ________ in which nutrients are transported about the body and in which ________ are excreted.

③Because of the high ________ of water, large changes in heat production can take place within the animal with very ________ alteration in body temperature.

④The animal obtains its water from three sources: ________ water, water present in its ________, and ________ water, this last being formed during metabolism by the oxidation of hydrogen-containing organic nutrients.

⑤Because of this great variation in water content, the composition of foods is often expressed on a ________ basis, which allows a more valid comparison of nutrient content.

⑥In both plants and animals, proteins are the ________ nitrogen-containing compounds. In plants, in which most of the protein is present as ________, the concentration is ________ in the young growing plant and falls as the plant ________.

(2) Ture or false

①The water contents of the animal body varies with age. ( )

②The newborn animal contains more water than in the mature fat animal. ( )

③An animal will die more rapidly if deprived of food than if deprived of water. ( )

④The main component of the DM of pasture grass is fat, and this is true of all plants and many seeds. ( )

⑤Vitamins are present in plants and animals in large amounts, and many of them are important as components of enzyme systems. ( )

⑥Calcium and phosphorus are the major inorganic components of animals, whereas potassium and silicon are the main inorganic elements in plants. ( )

(3) Answer the following questions according to the passage

①What are thefunctions of water in the body?

②Why do thelarge changes in heat production take place within the animal with very little alteration in body temperature?

③what are thesources of water obtained by animals?

④What isrelated to the water content of growing plants?

(4) Translation of the following sentences into Chinese

①The animal obtains its water from three sources: drinking water, water present in its food, and metabolic water, this last being formed during metabolism by the oxidation of hydrogen-containing organic nutrients.

②Because of this great variation in water content, the composition of foods is often expressed on a dry matter basis, which allows a more valid comparison of nutrient content.

③One of the main reasons for the difference between plants and animals is that, whereas the cell walls of plants consist of carbohydrates material, mainly cellulose, the walls of animal cells are composed almost entirely of lipid and protein.

④An important difference between plants and animals is that, whereas the former can synthesise all the vitamins they require for metabolism, animals cannot, or have very limited powers of synthesis, and are dependent upon an external supply.

课文A 动物及其饲料

水

动物机体中的水分随年龄而变化。新生动物机体含水量为750~800g/kg，但在成年动物机体中这一比例下降为500g/kg。维持机体中的水分对生物体的生命至关重要：如果剥夺水分，动物将会比被剥夺的食物更快的死去。水作为一种溶剂在机体内发挥作用，其中的营养物质被运送到机体内，而排泄物则被排出体外。许多由酶引起的化学反应发生在溶液中，并涉及水解。由于水的比热高，即便动物体内发生较大的热反应，但动物体温的变化却非常小。水也有很高的蒸发潜热，它从肺和皮肤蒸发并在调节体温上发挥更大的作用。

动物从三种来源获得水：饮用水、饲料中的水和代谢水，代谢水是在代谢过程中由含氢有机营养物质氧化而形成的。饲料中的含水量是不同的，精料中含水量低至60g/kg，而一些块根作物中含水量可超过900g/kg。由于水分含量的变化很大，饲料成分往往是基于干物质进行表达的，这使得对营养成分的比较更有效。

生长期植物的含水量与生长阶段有关，幼苗比老植物含水量更大。在温带气候条件下，获取饮用水通常不是问题，并且可以持续不断给动物供应水。没有证据表明，在正常情况下过量饮水是有害的，且动物通常会饮用它们所需要的水量。

干物质及其组成

饲料的干物质（DM）被方便地分为有机物和无机物，虽然在活的生物体中没有如此明显的区别。许多有机化合物都含有矿物元素作为结构成分。例如，蛋白质含有硫，许多脂类和碳水化合物含有磷。

牧草干物质的主要成分是碳水化合物，所有的植物和种子都是如此。像花生这样的含油种子含有大量的蛋白质和脂肪物质，是非常特别的。相比之下，动物机体内的碳水化合物含量非常低。植物和动物之间有差异的主要原因之一是，植物的细胞膜由碳水化合物物质（主要是纤维素）组成，而动物细胞的细胞膜几乎完全由脂肪和蛋白质组成。此外，植物储存能量的方式主要是碳水化合物，如淀粉和果聚糖，而动物储存能量的方式主要是脂肪。

动物机体的脂肪含量是不同的，与年龄有关，年龄大的动物所包含的脂肪多于年龄小的动物。活的植物的脂肪含量相对较低，如牧草，干物质含有脂肪40~50g/kg。

在植物和动物中，蛋白质是主要的含氮化合物。在植物中，大部分的蛋白质都是以酶的形式存在的，幼苗的蛋白质浓度很高，当植物成熟后蛋白质浓度降低。在动物中，肌肉、皮肤、毛发、羽毛、羊毛和指甲都主要由蛋白质组成。

与蛋白质一样，核酸也是含氮化合物，它们在所有生物体的蛋白质合成中起着基础性作用。它们还携带着活细胞的遗传信息。

植物和动物体内的有机酸包括柠檬酸、苹果酸、延胡索酸、琥珀酸和丙酮酸。虽然它们的含量很少，但它们在细胞的总体代谢中起着重要的媒介作用。其他有机酸是在瘤胃或青贮饲料中的发酵产物，包括乙酸、丙酸、丁酸和乳酸。

维生素在植物和动物中的含量非常少，许多维生素是酶系统的组成部分，具有很重要的作用。植物和动物的一个重要区别是，植物可以合成代谢所需的所有维生素，而动物不能，只能合成有限的所需维生素，其余的依赖于外部供应。

无机物包含除了碳、氢、氧和氮之外的植物和动物中的所有元素。钙和磷是动物中的主要无机成分，而钾和硅是植物中的主要无机元素。

Part Two Extensive Reading

1. Text B Concentrates, by-Products, and other Supplements for Dairy Cattle

Introduction

Most dairy cattle feed ingredients can be classified as either forages and roughages or concentrates. Generally, concentrates, many of which are grains, protein supplements, or by-product feeds, have less fiber and more digestible energy.

Reasons for feeding concentrates to dairy cattle

Balancing rations to supply needed nutrients not in forages. Because forages usually are a more economical source of nutrients than concentrates, the typical dairy cattle feeding program is built around the maximum use of forages. On most United States dairy farms the best available forages do not provide all the nutrients needed for the most profitable feeding. Accordingly, concentrates are fed to make up deficiencies. Usable energy is the nutrient required in largest amounts. Depending on price relationships between feed and milk, often it is not most profitable to feed sufficient energy for maximum milk production. Likewise, maximum rates of gain in growing dairy cattle are not desirable or most profitable.

With the typical dairy cattle operation in the United States, the practical approach is to feed sufficient concentrates to provide a usable energy level that results in the most profitable production. When this has been determined, the composition of the concentrate should be formulated to supply the other required nutrients not furnished by the forages. Mainly, concentrates are designed to balance the diet for the needed protein, minerals, and vitamins not in the forages.

Concentrate feeding (when concentrates are more economical sources of nutrients than forages)

Occasionally forages are in short supply or so expensive that concentrates are a more economical source of nutrients. In such situations, it is desirable to make the maximum effective use of them. Except for inadequacies in some needed nutrients, especially energy, feeding dairy cattle only forages does not cause problems. In contrast, serious digestive and health abnormalities often result when dairy cattle are fed only concentrates. Some of the problems associated with feeding excessive amounts of concentrates are due to insufficient intakes of unground fibrous feed.

With too high a proportion of concentrates, the overall digestibility of the ration is reduced. The highest, practical ratio of concentrates to forages, when they are more economical sources of nutrients than forages, depends on a number of variables. When it is fibrous, a smaller proportion of forage will suffice. Likewise, some concentrates such as beet pulp or citrus pulp contain appreciable fiber.

As a general rule, for best performance, lactating cows should not be fed more than about 60% of their total feed dry matter as concentrates; or less than about 40% as forage and roughages. Typically feeding more than 60% concentrates tends to decrease total milk yield, reduce fat percentage in the milk, and cause digestive and metabolic disturbances. In most situations, 17% crude fiber or 21% acid detergent fiber, from unground feeds, in the total ration dry matter is adequate even though the percentage of concentrates exceeds 60%.

Most of the above discussion concerns concentrate feeding for best performance. If the price relationships become distorted to a sufficient degree, it may be profitable to sacrifice some reduction in performance and accept some digestive problems, especially for short periods. Experimentally, cows have been fed rations containing 100% concentrates.

With excessive levels of concentrates, a number of additives, especially certain buffers, have at least partially alleviated the reduced milk fat percentage. Sodium and potassium bicarbonate and magnesium oxide appear to have some effect in such situations. Bentonite, a special type of clay which swells in water to many times its original size, also partially alleviates the low fat test exhibited with too high a proportion of concentrates. Apparently none of the buffers will increase milk fat percentage in cows fed a "normal" diet.

2. Notes to the Text B

(1) Depending on price relationships between feed and milk, often it is not most profitable to feed sufficient energy for maximum milk production.

句中"it is…"是强调句，"it"指代后面的"to feed sufficient energy for maximum milk production"。

(2) When this has been determined, the composition of the concentrate should be formulated to supply the other required nutrients not furnished by the forages.

句中"When"引导时间状语从句。"not furnished by the forages"为定语，修饰"the other required nutrients"。

(3) If the price relationships become distorted to a sufficient degree, it may be profitable to sacrifice some reduction in performance and accept some digestive problems, especially for short periods.

句中"If"引导条件状语从句。"it"指代"to sacrifice some reduction in performance and accept some digestive problems"。

3. Answer Questions

(1) How to classify the most dairy cattle feed ingredients?

(2) Why is the typical dairy cattle feeding program built around the maximum use of forages?

(3) What's problems associated with feeding excessive amounts of concentrates?

(4) What will happen if lactating cow is fed more than 60% concentrates?

(5) How to alleviate the low fat test exhibited with too high a proportion of concentrates?

课文B 奶牛的精料、副产物和其他饲料补充料

引言

大多数奶牛饲料可分为草料、粗饲料或精料。一般来说，精料中许多是谷物饲料、蛋白质补充料、副产物饲料，含有较少的纤维素和更易消化的能量。

饲喂奶牛精料的原因

草料中不含有全部的营养，但由于草料通常是比精料更经济的营养来源，所以典型的奶牛饲养计划是建立在最大限度使用草料的基础上。在大多数美国的奶牛场中，最好的草料也不能提供奶牛所有的营养需要。因此，需要饲喂奶牛精料以弥补不足，最大化地补充营养所需的可用能量。根据饲料和牛乳的价格关系，通常来说，提供充足能量使牛乳产量最大化并不是最赚钱的。同样，为生长期奶牛提供最大量的谷物也不是令人满意或利润最高的。

在美国典型的奶牛养殖场，实际的方法是提供足够的精料，以提供可用能量，从而保证生产盈利性最高。在确定了这一点后，应制定精料的组成，以供应草料中未提供的其他必需营养物质。主要的是，精料的设计是为了平衡在草料中不能提供的奶牛所需的蛋白质、矿物质和维生素。

饲喂精料（精料中的营养物质来源比草料更经济有效）

有时，草料供应短缺或价格过高，精料即是一种更经济的营养来源。在这种情况下，最有效地是利用精料来满足需要。草料除了缺少一些必需的营养物质，草料的能量也不足，但只饲喂奶牛草料并不会造成问题。相比之下，当奶牛只吃精料时，通常会导致严重的消化和健康问题。与饲喂过量的精料相关联的一些问题，通常是由于奶牛没有摄入足够的未研磨过的纤维性饲料。

由于精料的比例过高，整个日粮的消化率降低。当精料是比草料更经济的营养来源时，精料的实际比例最高，取决于许多变量。当它是纤维性的精料，补充一小部分草料就能满足奶牛需要。同样，一些精料如甜菜粕或柑橘渣也含有很多的纤维素。

一般来说，为了达到最佳生产性能，泌乳牛不应饲喂超过其总饲料量干物质60%的精料；或者低于40%的草料和粗饲料。一般情况下，饲喂奶牛60%以上的精料会降低牛乳的总产量，降低牛乳中的脂肪含量，导致消化和代谢紊乱。在大多数情况下，即使精料的百分比超过60%，在未研磨过的饲料中也有17%的粗纤维或21%的酸性洗涤纤维，在总的日粮干物质中是足

够的。

以上讨论的大部分关于饲喂精料的内容都是为了获得奶牛的最佳生产性能。如果价格关系扭曲到一定程度，牺牲一些生产性能的降低并接受一些消化问题，尤其是短期内的消化问题，可能是有益的。从实验上看，饲喂牛含100%精料的日粮利润是最高的。

过量的精料和一些添加剂，尤其是一些缓冲剂，至少能部分缓解牛乳乳脂率的减少。碳酸氢钠、碳酸氢钾和氧化镁在这种情况下似乎有一些效果。膨润土是一种特殊的黏土，它在水中能膨胀到原来体积的数倍，可部分缓解试验中精料比例过高导致的乳脂率降低的问题。显然，当饲喂“正常”日粮时，没有一种缓冲剂能增加牛乳中的乳脂率。

扫码进行拓展学习

Lesson 8
Feeding

Part One Intensive Reading

1. Learning Objective

After learning this lesson, you should study and understand the following:

(1) Give an overview of why feeding the lactating dairy cow is different from any other farm animal.

(2) Explain why most of the best practical feeding programs for milking herds are based on the use of large amounts of high quality forages.

(3) Explain why feeding young dairy calves during the first three months is a relatively small proportion of the total feed cost on a dairy farm.

2. Text A Feeding of Dairy Cattle

Feeding the milking herd

More than two-thirds of the total feed used on dairy farms is for the milking herd. In several respects, feeding the lactating dairy cow is different from any other farm animal. Good nutrition for the milking herd is an indispensible part of successful dairy farming, and a comprehensive overall plan to provide the needed nutrients from feeds that are sufficiently palatable to obtain adequate consumption is important. The feeds are needed every day of the year, and the cost must be reasonable if the operation is to be adequately profitable.

The total amount of feeds required for a moderate sized dairy herd during a year is huge. A typical 635 kg cow will eat about 20.4 kg of air-dry feed each day. For 100 cows milking herd this is about 2.25 tons daily or 820 tons per year. If a part of the feed is silage, pasture, or green chopped forage, the total weight involved is even greater.

Most of the best practical feeding programs for milking herds are based on the use of large amounts of high quality forages. Because high-producing cows usually cannot consume sufficient forage to fully meet their energy requirements, supplemental concentrates are needed. Since forages often are deficient in one or more other needed nutrients, the concentrate mixture should be formulated to make up any deficiencies.

Feeding and raising the young dairy calf

Dairy cows stay in the herd, on an average, for only about four lactations. When death and other losses of calves are considered, about three-fourths of the heifer calves born must be raised to maintain equal cow numbers. Raising all the heifer calves gives more opportunity for culling on the basis of desirable characteristics and generally is a very profitable practice. In most dairy areas there are several advantages to raising herd replacements rather than purchasing them. These advantages include more opportunity for genetic improvement, better diseases control, elimination of dependency on outside sources of cows with accompanying transportation and other added costs, and usually a lower cost for replacements.

Feeding young dairy calves during the first three months is a relatively small proportion of the total feed cost on a dairy farm. Because it does not have a functional rumen, the newborn calf has very different nutrient requirements from older dairy cattle. Its nutrient needs are more similar to those of simple stomached animals than to older cattle. The newborn calf is a baby which presents many nutritional and management problems including a much greater tendency to succumb to infections and other problems that may be fatal.

Feeding and management of heifers, dry cows, and bulls

Compared with baby calves and high-producing cows, heifers, dry cows, and bulls are relatively easy to feed. Perhaps this comparative ease is one reason these dairy animals often are neglected. Although the nutritional requirements of heifers, dry cows, and bulls are less demanding, the total amount of feed required is a substantial percentage of the total on a dairy farm.

3. New Words and Phrases

milking herd 挤奶牛群

herd [hɜːd] *n.* 兽群，畜群；放牧人；*vi.* 成群，聚在一起；*vt.* 放牧；使成群

lactating dairy cow 泌乳牛

farm animal 家畜

consumption [kən'sʌmpʃən] *n.* 消费；消耗；肺痨

profitable ['prɒfɪtəbəl] *adj.* 有利可图的；赚钱的；有益的

air-dry feed 风干饲料

deficiency [dɪ'fɪʃnsi] *n.* 缺陷，缺点；缺乏；不足的数额

dairy ['deərɪ] *n.* 乳制品；乳牛；制酪场；乳品店；牛乳及乳品业；*adj.* 乳品的；牛乳的；牛乳制的；产乳的

calf [kɑːf] *n.* 腓肠，小腿；小牛；小牛皮；（鲸等大哺乳动物的）幼崽

lactation [læk'teɪʃən] *n.* 哺乳；哺乳期；授乳（形容词 lactational）；分泌乳汁

heifer ['hefə] *n.* 小母牛

genetic improvement 基因改良

disease control 疾病控制

tendency ['tendənsɪ] *n.* 倾向，趋势；癖好

succumb [sə'kʌm] *vi.* 屈服；死；被压垮

infection [ɪn'fekʃən] *n.* 感染；传染；影响；传染病

fatal ['feɪtl] *adj.* 致命的；重大的；毁灭性的；命中注定的

substantial [səb'stænʃl] *adj.* 大量的；实质的；内容充实的；*n.* 本质；重要材料

4. Notes to the Text A

(1) These advantages include more opportunity for genetic improvement, better disease control, elimination of dependency on outside sources of cows with accompanying transportation and other added costs, and usually a lower cost for replacements.

句中“include”后面为并列的名词短语作宾语，均指代某种“advantage”。

(2) Compared with baby calves and high-producing cows, heifers, dry cows, and bulls are relatively easy to feed.

句中“Compared with…”表比较。“baby calves”指刚出生的小牛犊；“heifers”指小母牛；“dry cows”指不产乳的母牛；“bulls”指公牛，这些牛均不产乳。

5. Exercises

(1) Fill in the blanks by finishing the sentences according to the passage

①More than two-thirds of the total feed used on dairy farms is for the __________.

②Good nutrition for the milking herd is an ________ part of successful dairy farming, and a comprehensive overall plan to provide the needed ________ from feeds that are sufficiently ________ to obtain adequate consumption is important.

③If a part of the feed is ________, ________, or green chopped forage, the total weight involved is even greater.

④Since forages often are ________ in one or more other needed nutrients, the concentrate mixture should be formulated to make up any deficiencies.

⑤In most dairy areas there are several ________ to raising herd replacements rather than ________ them.

⑥Feeding young dairy calves during the first ________ is a relatively small proportion of the total feed cost on a dairy farm.

⑦ Compared with baby calves and high - producing cows, heifers, dry ________, and ________ are relatively easy to feed.

(2) Ture or false

①The total amount of feeds required for a moderate sized dairy herd during a year is huge. (　　)

②A typical 635 kg cow will eat about 20.4 kg of air-dry feed each day. (　　)

③Most of the best practical feeding programs for milking herds are based on the use of large amounts of low quality forages. (　　)

④Dairy cows stay in the herd, on an average, for only about two lactations. (　　)

⑤When death and other losses of calves are considered, about three-fourths of the heifer calves born must be raised to maintain equal cow numbers. (　　)

⑥Feeding young dairy calves during the first three months is a relatively large proportion of the total feed cost on a dairy farm. (　　)

⑦The newborn calf has the same nutrient requirements from older dairy cattle. (　　)

⑧Compared with baby calves and high-producing cows, heifers, dry cows, and bulls are relatively hard to feed. (　　)

(3) Answer the following questions according the passage

①How to feed the milking herd?

②How many heifer calves born must be raised to maintain equal cow numbers?

(4) Translation of the following sentences into Chinese

①Good nutrition for the milking herd is an indispensible part of successful dairy farming, and a comprehensive overall plan to provide the needed nutrients from feeds that are sufficiently palatable to obtain adequate consumption is important.

②Although the nutritional requirements of heifers, dry cows, and bulls are less demanding, the total amount of feed required is a substantial percentage of the total on a dairy farm.

③Even so, most culling is for low production, mastitis, lack of reproduction, disease problems, undesirable disposition, and other factors related to producing milk.

课文A 奶牛饲养

挤奶牛的饲养

奶牛场使用的饲料中，超过三分之二是饲喂挤奶牛的。在某些方面，饲喂泌乳牛与饲喂其他家畜不同。为挤奶牛提供良好的营养是奶牛养殖成功不可或缺的一部分，全面而整体的计划是从饲料中提供挤奶牛所需的营养物质，该饲料应适口性良好且足够挤奶牛消耗，这是很重要的。每天都需要这些饲料，如果要想获得足够的利润，成本必须合理。

一个中等体型的挤奶牛在一年里所需要的饲料总量是巨大的。一个标准的635kg的牛每天会吃掉约20.4kg的风干饲料。对于100头挤奶牛来说，每天约2.25t或每年约820t。如果饲料的一部分是青贮饲料、牧草或绿色切碎草料，饲料的总质量更大。

大多数挤奶牛的最佳饲喂规程是以使用大量高质量的草料为主。由于高产奶牛通常不能通过饲喂大量草料来满足其能量需要，所以需要额外补充精料。由于草料往往缺乏一种或多种其他必需营养物质，应配制精料以弥补其不足。

幼年小母牛的饲养

在牛群中的奶牛，平均只有4个泌乳期。考虑到会有犊牛死亡或其他损失时，必须饲养大约四分之三的小母牛来维持母牛数量。饲养所有的小母牛能提供更多的机会来根据需要挑选母牛，这通常是一种非常有益的做法。在大多数奶牛养殖区，养牛实现自然更替比购买好处多。这些优点包括有更多的机会进行基因改良，更好的疾病控制，消除对外来母牛的依赖，以及随之而来的运输和其他附加费用，而且通常替换成本较低。

在头三个月里，饲喂幼年小母牛只占奶牛场饲料总成本的一小部分。因为它没有功能性瘤胃，所以新生的小牛犊和大龄的奶牛营养需要不同。它的营养需要更类似于那些简单的单胃动物，而不是大龄牛。初生牛犊更像是一个婴儿，表现出许多营养和管理问题，包括更容易受到感染和其他可能致命的问题。

小母牛、干奶牛和公牛的饲喂管理

与幼年小牛犊和高产奶牛相比，小母牛、干奶牛和公牛都相对容易饲喂。也许这正是这些奶牛经常被忽视的原因之一。虽然小母牛、干奶牛和公牛的营养需要较低，但所需饲料的总量在奶牛场中占相当大的比例。

Part Two Extensive Reading

1. Text B Feeding of Pigs

Feeding the baby pig

Feeding of pigs has advanced to the point where the growth period is divided into several segments, and the pig is fed according to the stage of the growth cycle it is in. The younger the pig, the more critical the period becomes. Consequently, higher quality and more highly fortified diets are needed.

The trend is toward early weaning of pigs at 3~5 weeks of age so sows can be maximize the number of litters obtained yearly. Thus, prestarter diets and starter diets are used, depending on age of weaning, milking ability of sows, litter size and other factors.

The diet which the sow receives during its growing period, as well as during gestation, will definitely affect the ability of the baby pigs to survive and grow after weaning. This means that gilts kept for reproduction must be fed well-balanced diets during their growing period to develop the normal, reproductive tract needed for successful production of large litters. A high quality diet must be fed to the sow during gestation to ensure proper growth and development of the fetus. One of the biggest mistakes pigs producers make is to start feeding a sow well-balanced diets only 3 or 4 weeks before farrowing. They have heard that the developing fetuses grow most in this period (which is true) . As a result, they feed the sow a well-balanced diet only then. Until that time, they may feed the sow a poor quality diet in an effort to save on feed costs. As a result, the sow does not obtain the nutrients needed and so farrows many weak or dead pigs and ends up weaning small, poor-looking litters. The answer to this problem is never to deprive the sow of needed protein, minerals, and vitamins. Any lack of these nutrients in the diet means a loss of pigs, and these lost pigs are never marketed. Presently, 35% ~ 45% of the ovulated eggs fail to be represented by live pigs at birth and 20% ~ 25% of the live pigs at birth die before weaning. This is a tremendous loss which must be eliminated.

The heavier the pig is at birth, the heavier it will be at weaning. On the average, a difference of 0.45 kg in weight at birth was accompanied by 3.53 kg difference at weaning. This information shows the value of properly feeding the sow during gestation so that it will farrow heavy, thrifty pigs. Such pigs not only will have a better chance to survive but also will be heavier at weaning.

The pigs which are heavier at weaning also tend to do better afterwards. A comprehensive study at the Minnesota Station involving 745 gilt litters, composed of 5562 pigs born alive, of which 3918 pigs survived to weaning also showed that the average birth weight had the greatest effect, of the factors studied, upon both survival and total weaning weight of the litters.

An early attempt to feed a synthetic milk and thus wean pigs right after birth was unsuccessful because sucrose was fed as the source of carbohydrate. Later, Illinois workers showed that the newborn pig cannot utilize sucrose; this partly accounted for the early failure. Shortly after the development of synthetic milks in 1951, pig hatcheries sprang up in many sections of the country. These pig hatchers were establishments which specialized in producing and selling weanling pigs throughout the year for growing and finishing on the farms of purchasers. The hatcheries ranged in scale of operation from 40 or 50 sows on up to 400 or more.

The pig hatchery, however, was characterized by many failures. Disease, lack of sanitation, and poor management were the chief causes of the failures. Synthetic milks were not widely accepted. Liquid milk often spilled over and soaked the litter on the floor which chilled the pigs, caused scouring, etc. Also, keeping the equipment clean and sanitary provided many problems; many cases of scours and deaths resulted from unsanitary equipment. Souring of the milk in the feeders was particularly trouble some when the pigs did not promptly eat their feed. As a result, only a few of the hatcheries, those with qualified personnel and with good management procedures, were able to raise pigs successfully under this system.

2. Notes to the Text B

（1）Thus，prestarter diets and starter diets are used，depending on age of weaning，milking ability of sows，litter size and other factors.

句中“starter diets”意为开食料。

（2）Souring of the milk in the feeders was particularly trouble some when the pigs did not promptly eat their feed.

前一句中的“scouring”意为腹泻，本句中的“souring”意为变酸。两个单词很像，注意区分。

3. Answer Questions

（1）How to select prestarter diets and starter diets?

（2）How to feed sow?

（3）How many kilograms difference in weight at weaning was accompanied by a difference of 0.45 kg in weight at birth on the average?

（4）What did Illinois workers show?

（5）Why was synthetic milk not widely accepted?

课文B　猪的饲养

仔猪饲养

猪是按生长期分成几个阶段进行饲喂的。猪越小，其生长周期越重要。因此，需要更高质量和更严格的日粮。

这种趋势存在于3~5周龄的早期断乳仔猪中，这样可使每年得到的仔猪数达到最大。因此，应根据断乳日龄、母猪泌乳能力、产仔数等因素，选择预开食料或开食料。

母猪在生长期和妊娠期所采食的日粮，一定会影响仔猪在断乳后的存活和生长能力。这就意味着，在生长期，为了成功生产大量的仔猪，必须饲喂母猪营养均衡的日粮，使其生殖系统发育正常。为了确保胎儿的正常生长和发育，必须在妊娠期饲喂母猪高质量的日粮。养猪生产者最大的错误之一就是仅在母猪分娩前3周或4周才开始饲喂母猪营养均衡的日粮。他们认为发育中的胎儿在这个时期生长得最多（这是真的）。因此他们只在这个时期饲喂母猪营养均衡的日粮。在此之前，他们可能会饲喂母猪低质量的日粮，以节省饲料成本。结果，母猪没有获得所需营养，产下许多瘦弱或死亡的猪仔，最后断乳的仔猪又小又弱。这个问题的答案是，永远不要剥夺母猪所需的蛋白质、矿物质和维生素。在日粮中缺少这些营养素意味着猪的损失，而这些被损失掉的猪没有机会推向市场。目前，生猪生产中有35%~45%的排卵失败，20%~25%的生猪在断乳前死亡。这是一个巨大的损失，必须尽力消除。

猪出生时越重，断乳时就越重。平均而言，出生时体重差异0.45kg，在断乳时体重差异为3.53kg。这一信息显示了在妊娠期正确饲喂母猪的价值，这样它就可以产出又大又好的仔猪。这样的猪不仅有更好的存活率，而且在断乳时体重更重。

在断乳时体重较重的猪以后的体重也更重。明尼苏达州的一项综合研究涉及了745头母猪，其中出生的仔猪有5562头存活，并有3918头猪存活到断乳期，这也显示出在研究的多种因素中，平均出生重的影响最大，对仔猪存活率和断乳重都有影响。

由于蔗糖是合成乳中碳水化合物的来源，所以早期尝试在仔猪刚出生时就饲喂合成乳，从而使猪断乳。后来，伊利诺伊州的员工发现，新生仔猪不能利用蔗糖，这在一定程度上解释了早期断乳失败的原因。在合成乳发展不久之后，1951年，在中国的许多地区都出现了仔猪孵育场。这些养猪生产者是专门从事生产和销售断乳仔猪的机构，他们在这一年里不断地生长和育成断乳仔猪。这些孵育场的生产规模从40头或50头母猪到400头以上母猪。

然而，猪繁育场还是有许多失败。疾病、卫生条件缺乏和管理不善是导致这些失败的主要原因。合成乳没有被广泛应用。液体乳经常外溢，把地板上的垫草弄湿，使猪受冻，造成腹泻等。同时在保持设备干净卫生方面带来了许多问题，因设备不卫生造成许多腹泻和死亡病例。当猪没有及时吃完饲料时，供料设备里的乳变酸尤其麻烦。因此，只有部分繁育场，拥有合格的员工和良好的管理规程，能够成功养猪。

扫码进行拓展学习

Lesson 9

Pasture

Part One Intensive Reading

1. Learning Objective

After learning this lesson, you should study and understand the following:

(1) Explain how many ways pastures are used in.

(2) Describe factors are related to the conception rate.

(3) Give an overview of which supplementation is needed.

2. Text A Good Horses with Good Pastures

Good horses and good pastures go together. By instinct, horses are forage consumers. They are also particular about the pasture plants they like to eat. Pastures are not only a source of excellent-quality feed but also provide a means of exercise and help in bone and body development. Most horse people feel there is no substitute for a high-quality pasture. Many of them feel the other feeds used are a supplement to the pasture rather than vice versa. Good pastures are excellent sources of protein, vitamins, minerals, and other nutrients. Moreover, they save considerably on the feed bill. So, a good horse-breeding farm program should be built around an excellent-quality pasture program. Pastures are ideal for young animals and breeding stock. Pastures are used in many ways. Three of the most important are as follows: (1) to provide all or most of the feed; (2) to provide a large portion of feed needs, the remainder coming from supplements and possibly some energy source; (3) to provide very little feed but to be used primarily as a holding area or for exercise purposes.

Unfortunately, what many horse people use as a pasture is more of an exercise area rather than a source of feed. Therefore, if pastures are going to provide a significant amount of the feed supply, they need to be high quality and managed properly. It is becoming increasingly more difficult to have a good pasture program in suburban areas where the greatest numbers of horses exist since there is a lack of sufficient land available for this purpose.

Pastures are ideal for young animals and breeding stock. Many horse owners who have difficulty getting their mares to breed, have observed that a lush, green pasture is beneficial in getting them to conceive. In some areas, the conception rate does not start to improve until pastures grow and become green in the spring. This is one reason why the greatest conception rates occur in March, April, and May. Whether something in the pasture causes the increase in conception rate or whether some other factors are involved is not definitely known. It is possible that some environmental factor (temperature, sun's rays, length of daylight, humidity, etc.) could be involved. Or, the pasture could be supplying some nutrients that increase conception rate. But regardless of what is involved, most horse owners prefer to have as much high-quality pasture as possible. Unfortunately, many of the top horse farms producing high-level performance and race horses are located near large cities where land is scarce and high priced and many have to settle for a limited pasture acreage. In some areas, the weather is such that pastures cannot be raised during the cold months which also limits the amount of pasture available during the year. The next alternative is to use high-quality hay and/or high-quality dehydrated alfalfa meal as substitutes for limited pasture availability. Some horse farms have put in temporary pastures such as oats, rye, wheat, barley, or others for green grazing during the winter months. In certain localities, irrigation is needed for these temporary pastures to do well. Since horses have a birthday each January 1, it is important to have them foaled as close to this date as possible. Therefore, the use of green pasture at breeding time can be of help in getting mares bred earlier.

A good-quality pasture should be kept lush, green, and not allowed to get too short or too tall. If allowed to get too short, the horses will eat the pasture too close to the ground. This not only increases the danger of parasite infestation, but also may be harmful to the pasture since the horses can overgraze it too close to the roots. If the pasture gets too tall, it becomes too mature and decreases in digestibility and feeding value. For good parasite prevention and good pasture management, therefore, the horses should be rotated between pastures periodically in order to prevent overgrazing, to rest each field regularly, and to allow it to recover from grazing.

In many cases, pastures supply only part of the total feed needs of the horse. This is especially the case with high-level performance horses that are exercising, growing, or producing a foal. Texas A & M showed that yearling stock horses on well-managed bermudagrass and ryegrass pastures plus trace mineralized salt can attain similar or greater average daily gains than projected by the 1978 NRC report. How much supplementation is needed therefore depends on many factors which include the following: (1) quality and quantity of pasture; (2) the condition of the horses; (3) how much is demanded of the horse in work or performance; (4) the stage of the life cycle of the horse (such as foal, yearling, 2 years old, mare, or stallion); (5) how well the owner wants the horses to look; and (6) many others.

Forages (pasture and hay) make up a large portion of horse diets. Horses by instinct are forage consumers. They should be fed at least 1.0% of their body weight as forage (on a dry matter basis). This means 4.54 kg of hay, or other forage equivalent, for a 454 kg horse. Some people feed forage at a higher level and some at a lower level. It is thought that the minimum forage requirement is 0.5% of body weight or 2.3 kg per 454 kg horse. But, many prefer to feed forage at a level of at least 1.0% of body weight.

It is important to maximize forage utilization. Information on digestibility of forage is important since it provides a guideline on when to graze or when to make hay from excess forage. Most of the present data on digestibility of forages are available with hay. It is reasonable to assume, however, that the information obtained with hay would have considerable application to the pasture from which the hay was made. Therefore, some experimental data with hay will be presented here.

The pastures to use will vary with the geographical area of the country. The County Agent, Farm Advisor, or State University can be of considerable help in advising on the pasture programs that are best for the area. Legumes such as alfalfa, ladino clover, red clover, birdsfoot trefoil, and others are excellent and high in nutritional value. Sometimes legumes are used alone, but usually they are mixed with grasses. Grasses such as timothy, bromegrass, orchardgrass, canary grass, bermudagrass, bluegrass, buffalograss, crested wheatgrass, and others are used extensively with horses. Temporary pastures such as oats, wheat, rye, barley, and others are also used to supplement the permanent pastures. If properly managed and cared for, many kinds of pastures can be used successfully.

3. New Words and Phrases

forage [ˈfɒrɪdʒ] *n.* 饲料；草料；搜索；*vi.* 搜寻粮草；搜寻

bone [bəʊn] *n.* 骨；骨骼；*vt.* 剔去……的骨；施骨肥于；*vi.* 苦学；专心致志

supplement [ˈsʌplɪmənt] *vt.* 增补，补充；*n.* 增补，补充；补充物；增刊，副刊

horse-breeding farm 繁育马场

excellent-quality 优质的

breeding stock 种畜

holding area 驯养区

suburban area 郊区

conceive [kənˈsiːv] *vt.* 怀孕；构思；以为；持有；*vi.* 怀孕；设想；考虑

conception rate 受孕率

environmental factor 环境因素

temperature [temprətʃər] *n.* 温度；量某人的体温；发烧；氛围；体温

sun's ray 太阳光

length of daylight 日照时间

humidity [hjʊˈmɪdətɪ] *n.* 湿度；湿气

acreage [ˈeɪkərɪdʒ] *n.* 面积，英亩数

hay [heɪ] *n.* 待割的草；干草

dehydrated alfalfa meal 脱水苜蓿粉

availability [əˌveɪləˈbɪləti] *n.* 可用性；有效性；实用性

oat [əʊt] *n.* 燕麦；麦片粥，燕麦粥

rye [raɪ] *n.* 黑麦；吉卜赛绅士；*adj.* 用黑麦制成的

wheat [wiːt] *n.* 小麦；小麦色

barley [ˈbɑːrli] *n.* 大麦

grazing [ˈgreɪzɪŋ] *n.* 放牧；牧草；*v.* 擦过；抓伤（graze 的现在分词）

irrigation [ˌɪrɪˈgeʃn] *n.* 灌溉；冲洗；冲洗法

lush [lʌʃ] *adj.* 丰富的，豪华的；苍翠繁茂的；*vi.* 喝酒；*n.* 酒；酒鬼；*vt.* 饮

parasite [ˈpærəsaɪt] *n.* 寄生虫；食客

infestation [ˌɪnfɛˈsteɪʃən] *n.* 感染；侵扰

overgraze [ˌovərˈgrez] *vi.* 过度放牧；*vt.* 在……上过度放牧

mature [məˈtʃʊr] *vi.* 成熟；*adj.* 成熟的

digestibility [daɪdʒɛstəˈbɪləti] *n.* 消化性；可消化性；消化率

feeding value 饲喂价值

rotate [rəʊˈteɪt] *vi.* 旋转；循环；*vt.* 使旋转；使转动；使轮流；*adj.* [植]

辐状的

periodically [ˌpɪərɪˈɒdɪkəlɪ] *adv.* 定期地；周期性地；偶尔；间歇

foal [fəʊl] *n.* 驹（尤指一岁以下的马、驴、骡）；*vi.*（马等）生仔；*vt.*（马等）生仔

yearling [ˈjɪəlɪŋ] *n.* 一岁家畜；满一岁的动物；*adj.* 一岁的

bermudagrass [bəˌmjʊdəˈgræs] 狗牙根草

ryegrass [ˈraɪgræs] *n.* 黑麦草

trace mineralized salt 微量矿化盐

average daily gain 平均日增重

quality [ˈkwɒlətɪ] *n.* 质量，品质；特性；才能；复数 qualities

quantity [ˈkwɒntɪtɪ] *n.* 量，数量；大量；总量；复数 quantities

stallion [ˈstæliən] *n.* 种马；成年公马

requirement [rɪˈkwaɪəmənt] *n.* 要求；必要条件；必需品

body weight 体重

geographical [dʒɪəˈgræfɪkəl] *adj.* 地理的；地理学的

county agent 农区指导员

farm advisor 农场顾问

state university 州立大学

legume [ˈlegjuːm] *n.* 豆类；豆科植物；豆荚

alfalfa [ælˈfælfə] *n.* 苜蓿；紫花苜蓿

ladino clover 白三叶草

red clover 红三叶草

birdsfoot trefoil 百脉根

nutritional value 营养价值

grass [grɑːs] *n.* 草；草地，草坪；*vt.* 放牧；使……长满草；使……吃草；*vi.* 长草

timothy [ˈtɪməθɪ] *n.* 梯牧草

bromegrass [ˈbromgræs] *n.* 雀麦草；雀麦属植物

orchardgrass [ˈɔːrtʃərˈgræs] *n.* 野茅；果园草

canary grass 加那利草

bluegrass [ˈbluːgræs] *n.* 莓系属的牧草；早熟禾属植物

buffalograss 野牛草

crested wheatgrass 冠毛大麦草

4. Notes to the Text A

(1) It is becoming increasingly more difficult to have a good pasture program in suburban areas where the greatest numbers of horses exist since there is a lack of sufficient land available for this purpose.

句中“It is…”是强调句，强调形容词“more difficult”，“it”指代后面的“to have a good pasture program in suburban areas”。“where”引导定语从句，修饰先行词“suburban areas”。“Since”引导原因状语从句。

(2) But regardless of what is involved, most horse owners prefer to have as much high-quality pasture as possible.

句中“regardless of”意为不管，不顾，表让步。“as much…as possible”意为尽可能多。“much”后接不可数名词，如后接可数名词，此处“much”换为“many”。

(3) This not only increases the danger of parasite infestation, but also may be harmful to the pasture since the horses can overgraze it too close to the roots.

句中“not only…, but also…”意为不仅……，而且……。“be harmful to…”意为对……有害。“since”引导原因状语从句。“too+形容词/副词+to…”译为“太……而不能……”，它在形式上是肯定的，但在意义上是否定的。

(4) Since horses have a birthday each January 1, it is important to have them foaled as close to this date as possible.

本句直译是“由于马的生日是每年的1月1日，所以要尽可能在这个日期前生仔”。在美国，如果家里有马，一般都要参加俱乐部，进行登记，不论是什么时间出生，都登记为当年的1月1日，竞赛时以马的年龄分组。

5. Exercises

(1) Fill in the blanks by finishing the sentences according to the passage

①Pastures are not only a ________ of excellent-quality feed but also provide a means of exercise and help in ________ and ________ development.

②Good pastures are an excellent source of ________ , minerals, and other nutrients.

③Therefore, if pastures are going to provide a ________ amount of the feed supply, they need to be high ________ and managed properly.

④Pastures are ________ for young animals and breeding stock.

⑤In some areas, the ________ does not start to improve until pastures grow and become green in the spring.

⑥It is possible that some environmental factor (________ , sun's rays, length of daylight, humidity, etc.) could be involved.

⑦Since horses have a birthday each January 1, it is important to have them foaled as ________ to this date as possible.

⑧A good-quality pasture should be kept ________ , and not allowed to get too short or too tall.

⑨This not only increases the danger of parasite infestation, but also may be harmful to the pasture since the horses can overgraze it too close to the ________ .

⑩Forages (________ and ________) make up a large portion of horse diets.

(2) Ture or false

①Horses are not particular about the pasture plants they like and eat. ( )

②Good pastures are an excellent source of fat, protein, vitamins, minerals, and other nutrients. ( )

③Pastures are used to provide a large portion of feed needs, the remainder coming from supplements and possibly some energy source. ( )

④Texas A & M showed that yearling stock horses on well - managed bermudagrass and ryegrass pastures plus trace mineralized salt can attain similar or greater average daily gains than projected by the 1978 NRC report. ( )

⑤Forages (pasture and hay) should be fed at least 0.5% of their body weight as forage (on a dry matter basis). ( )

⑥Information on digestibility of forage is important since it provides a guideline on when to graze or when to make hay from excess forage. ( )

⑦Most of the present data on digestibility of forages are available with pasture. ( )

⑧If properly managed and cared for, many kinds of pastures can be used successfully. ( )

(3) Answer the following questions according to the passage

①Which ways are pastures used importantly?

②Why is difficult to have a good pasture program in suburban areas?

③Which factors are related to the conception rate?

④How to solve the problem that pastures cannot be raised during the cold months?

⑤How to use legumes such as alfalfa, ladino clover, red clover, birdsfoot trefoil, and others are excellent and high in nutritional value to feed horse?

（4） Translation of the following sentences into Chinese

①It is becoming increasingly more difficult to have a good pasture program in suburban areas where the greatest numbers of horses exist since there is a lack of sufficient land available for this purpose.

②Many horse owners who have difficulty getting their mares to breed，have observed that a lush，green pasture is beneficial in getting them to conceive.

③Unfortunately，many of the top horse farms producing high-level performance and race horses are located near large cities where land is scarce and high priced and many have to settle for a limited pasture acreage.

④For good parasite prevention and good pasture management，therefore，the horses should be rotated between pastures periodically in order to prevent overgrazing，to rest each field regularly，and to allow it to recover from grazing.

⑤It is important to maximize forage utilization. Information on digestibility of forage is important since it provides a guideline on when to graze or when to make hay from excess forage.

课文A　好马须有好牧场

好马通常有好的放牧牧场。吃草是马的天性，它们对自己喜欢吃的牧草也很讲究。牧场不仅是优质饲料的来源，而且是锻炼和帮助马骨骼和身体发育的一种方式。大多数养马人认为没有什么可以替代高质量的牧场，他们中的许多人认为其他饲料只是牧草的补充料。好的牧草是蛋白质、维生素、矿物质和其他营养物质的极佳来源。此外，饲喂牧草节省了大量的饲料费用。因此，一个好的马场计划应该围绕一个优质牧场计划来建立。牧场是青年动物和种畜的理想场所。牧场有很多用途，其中最重要的三个是：（1）提供全部或大部分饲料；（2）提供大部分饲料，其余部分来自补充料和可能还有一些能量来源；（3）提供非常少的饲料，但主要用于驯养区或锻炼目的。

不幸的是，许多养马人把牧场更多的是作为锻炼的场所，而不是饲料来源。因此，如果牧场将提供大量的饲料供给，这些牧草必须是高质量的，且经过正确管理的。由于缺乏足够的土地用于这一目的，在马数量最多的郊区建立一个良好牧场的计划变得越来越困难。

牧场是青年动物和种畜的理想场所。许多马主很难让他们的母马繁殖，他们发现茂盛的、绿色的牧场有利于它们受孕。在一些地区，直到春天牧草生长并变绿，受孕率才开始提高。这就是为什么3月、4月和5月是受孕率最高的月份。牧草里的某些物质是否会导致受孕率增加，或者是否涉及其他因素，目前还不清楚。可能与环境因素（温度、太阳光、日照时间、湿度等）有关。或者是因为牧草可以提供一些营养物质来提高受孕率。但不管涉及什么，大多数马主都希望拥有尽可能多的优质牧草。不幸的是，许多生产高水平性能的马匹和赛马的顶级马场都坐落在大城市附近，那里的土地稀少，价格昂贵，许多人不得不接受有限的牧场面积。在一些地区，由于天气原因，在寒冷季节不能种植牧草，这也限制了全年的牧草数量。另一种选择是使用高质量的干草和/或高质量脱水苜蓿粉作为有限牧草的有效替代品。有些马场在冬季种植临时牧草（如燕麦、黑麦、小麦、大麦等）替代绿色牧草。在某些地方，这些临时的牧场需要灌溉。由于马的生日是每年的1月1日，所以要尽可能在这个日期前产仔。因此，在母马繁殖期使用绿色牧草有助于母马较早繁殖。

一个好的牧草地应该保持茂盛嫩绿，不能长得太矮或太高。如果牧草太矮，马就会吃太靠近地面的牧草。这不仅增加了寄生虫感染的危险，而且可能对牧场有害，因为马可能会在离牧草根部太近的地方过度放牧。如果牧草太高，就会变得太成熟，消化率和饲喂价值就会降低。因此，为了很好地防治寄生虫和管理牧场，马应该定期在各牧场之间轮换，以防止过度放牧，定期让每一块牧场休牧，并使其恢复。

在许多情况下，牧草只提供马全部饲料需要的一部分。当高水平性能的马正在训练、生长或繁殖小马驹时，这种情况在它们身上尤为明显。Texas A & M研究表明，饲喂一岁的骑乘马管理良好的百慕大草、黑麦草加入微量矿化盐可以获得比1978年NRC报告相似或更大的平均日增重。因此需要补充多少牧草取决于许多因素，包括以下几点：（1）牧场的数量和质量；（2）马匹的体况；（3）马在训练或表演中的营养需要是多少；（4）马所处的生命周期阶段（如马驹、一岁马、两岁马、母马或种马）；（5）主人想让马看起来有多健壮；（6）其他。

草料（牧草和干草）是马日粮的主要组成部分。马是本能的草料消耗者。饲喂马的草料质量至少应该是马体重的1.0%（干物质基础）。这意味着454kg的马需要4.5kg干草，或者其他等量草料。有些人饲喂马较高水平的草料，有些人饲喂马较低水平的草料。据认为，最低的草料需要是体重的0.5%或每454kg马饲喂2.3kg草料。但是，许多人喜欢饲喂马至少体重1.0%的草料。

最大限度地利用草料是很重要的。有关草料消化率的信息是很重要的，因为它提供了何时放牧或何时利用过量草料制作干草的指南。目前有关草料消化率的大部分数据都是从干草中获得的。然而，合理的假设是，从干草中获得的信息将对生产干草的牧场很有价值。因此，本文将给出一些干草实验数据。

使用的牧场会根据国家地理区域的不同而有所不同。农区指导员、农场顾问或州立大学可以为该地区的牧场项目提供建议，对该地区有相当大的帮助。像苜蓿、白三叶草、红三叶草、百脉根等豆科植物都很好，营养价值很高。有时豆科植物被单独使用，但通常它们与禾本科植物混合使用。禾本科植物如梯牧草、雀麦草、果园草、加那利草、狗牙根草、莓系属牧草、野牛草、冠毛犬麦草等广泛用于饲喂马。临时种植的牧草如燕麦、小麦、黑麦、大麦等也被用来补充永久性牧草的不足。如果管理得当，许多牧草都可以被成功利用。

Part Two Extensive Reading

1. Text B Pasture for Horses

Managing pastures

Usually, horses will consume the immature, short pasture in preference to the more mature pasture. This can result in horses concentrating in one area and keeping it grazed closely while other areas grow untouched. This can be counteracted by dividing pastures into smaller areas and stocking the pastures with enough horses to keep the pastures short, lush, and green. One needs to guard against too many horses in a pasture, which can result in overgrazing and aggravation of the parasite problem.

Pastures require proper fertilization and management in order to keep them highly productive, nutritious, and palatable. Soil tests should be made periodically to determine the level of major and minor mineral elements, as well as pH (acidity), in the soil. These tests should be properly evaluated by the farm operator or by soil and pasture specialists. The advice of the County Agent or Farm Advisor is very helpful since they are acquainted with fertilizer needs in the county or area. The timing of the fertilizer applications can be very helpful in regulating the amount of forage available for grazing. The fertilizer formulation is also very important. Factors that are important in determining fertilizer needs include the following: (1) soil type; (2) kind of forages used; (3) soil fertility; (4) moisture level; (5) season of the year; (6) plant maturity; (7) plant health; and (8) climate as well as other factors.

Grass, legume, or grass-legume pastures can be used for horses. During certain periods of the year, some pastures may be too laxative for horses that are exercised a great deal. The laxative effect can be alleviated by feeding a dry grass hay, in addition to the pasture, or by letting the horses run into another field which has dried pasture. Brood mares and young colts should best be kept in pastures that give a laxative effect. The horses used for performance or racing purposes should be kept on pastures that keep the horses in good condition as far as their bowels and sweating are concerned. If necessary, one can also restrict the horses to a certain number of hours daily on the pasture in order to prevent too much of a laxative effect with heavily exercised or worked horses.

Many horse owners turn their animals out on pasture at night as well as on idle days. This gives the horses an opportunity to relax and exercise at will. It also decreases the amount of grain and hay that is fed.

It is important to avoid having stumps, pits, poles, holes, and any sharp obstacles in the pastures that can injure the horses. Many good horses have been ruined by these hazards in pastures. It is helpful to place mineral boxes in a corner of the pasture to avoid horses running into them. A plastic rather than wooden mineral box will also lessen the possibility of an injury.

2. Notes to the Text B

This can be counteracted by dividing pastures into smaller areas and stocking the pastures with enough horses to keep the pastures short, lush, and green.

句中“动词+ing”形式做介词宾语。

3. Answer Questions

(1) Which factors are important in determining fertilizer needs?

(2) Which management hints are given in the text for use in developing and maintaining quality pastures?

(3) How to alleviate the laxative effect of pastures?

(4) How to avoid sharp obstacles injure the horses in the pastures?

(5) Will a plastic rather than wooden mineral box lessen the possibility of an injury?

课文B 马的放牧

管理牧场

通常情况下，马更喜欢吃嫩的、矮的牧草而不是更成熟的牧草。这可能导致不均衡、马集中在一个区域，而不去其他区域。这可以将牧场分割成较小的区域并在牧场上为足够的马储备牧草来保持牧草的矮短、茂盛和嫩绿。需要防止在一片牧场上放牧太多的马，因为这会导致过度放牧和寄生虫问题的加剧。

牧场需要适当的施肥和管理，以保持其高产、营养和美味。应定期进行土壤试验，以确定土壤中常量元素和微量矿物元素水平以及土壤的pH（酸度）。这些试验应由农场经营者或土壤和牧场专家进行适当评估。农区指导员或农场顾问的建议很有帮助，因为他们了解农村或地区的肥料需求。施肥的时间有助于调控可供放牧的牧草数量。肥料配方也很重要。决定肥料需求的重要因素包括：（1）土壤类型；（2）使用的饲草种类；（3）土壤肥力；（4）湿度；（5）季节；（6）植物成熟度；（7）植物健康状况；（8）气候以及其他因素。

马可以利用禾本科牧草、豆科牧草或禾本科–豆科混合牧草。在一年中的某些时期，一些牧草对经常运动的马来说可能通便作用过强。除了牧草之外，还可以通过饲喂干草来减轻腹泻，或者让马跑到另一块已经干燥的牧场上放牧。种母马和小马驹最好养在有通便作用的牧场上。就马的肠胃和出汗而言，用于表演或比赛的马应放牧在能使马保持良好体况的牧场上。如果必要，人们也可以限制马每天在牧场上放牧的时间，防止对大量训练或工作的马产生过多的通便作用。

许多马主在晚上和空闲的日子里都把他们的马赶出牧场，给马一个放松和随意运动的机会，也减少了谷物和干草的饲喂量。

重要的是要避免在牧场上有树桩、坑、电线杆、洞和任何可能伤害马的尖锐障碍物。许多好马都被牧场上的这些危险给毁了。把矿物盒放在牧场的一个角落里有助于避免马撞到它们。一个塑料的而不是木制的矿物盒也会减少马受伤的可能性。

扫码进行拓展学习

Unit Ⅲ Management

Lesson 10
Restraint

Part One Intensive Reading

1. Learning Objective

After learning this lesson, you should understand the following:

(1) Give an overview of some degree of control over an animal's movement and activity is required for every technique that the livestock producer performs.

(2) restraint practices became important thousands of years ago.

(3) Animals are affected by our production systems and our management techniques.

(4) Explain why you should think of some questions when it is your responsibility to select the restraint method to be used in performing a given management technique.

(5) Give an overview of why it is the manager's responsibility to become proficient at it whatever the method of restraint selected.

(6) Describe the psychological principles underlying physical restraint techniques.

2. Text A Control over Animals

Some degree of control over an animal's movement and activity is required for every technique that the livestock producer performs. Restraint can vary from the psychological control that a handler's voice may exert his animals to the complete restriction of activity and total immobilization that chemical agents can provide. With large, potentially dangerous animal, a combination of psychological, physical, and chemical restraint is often employed.

Restraint practices became important thousands of years ago when man first domesticated animals for food and fiber and as beasts of burden. This domestication altered the natural lifestyle of this animals, and forced man to be responsible for the animal's needs. Managing these animals for human purposes necessitated control of the beasts; there is evidence of early restraint in crude fencelike enclosures. Not until many years later was there any concern about the appropriateness of the restraint for the task at hand.

Today, we realize that it is our responsibility, since we have total control over the animal's life, to be concerned for its welfare, its sensation of pain, and its psychological well-being as they are affected by our production systems and our management techniques. However, we should not overemphasized the sentiment of not causing the animal any discomfort and lose sight of the fact that animals are housed, maintained, and ministered to for the production of food, fiber, and pleasure. Certain management techniques must be performed to achieve those ends. Some of these techniques will cause a degree of pain.

Pain is a necessary phenomenon of nature that signals to the animal that something out of the ordinary is happening to its body. Without this sensation of pain, noxious factors could destroy an animal's body without his knowing it. The animal manager's responsibility is to perform his tasks of caring for the animals in the most currently appropriate manner, while inflicting the least amount of pain and causing the least amount of psychological upset (fright). When the animal is restrained and the manager is performing necessary management techniques, there is no escape from the pain for the animal, no relief from the fright. Anyone failing to realize this or failing to do everything within his power to alleviate the pain and fright should not be allowed to manipulate the lives of animals.

When it is your responsibility to select the restraint method to be used in performing a given management technique, you must ask yourself the following questions: (1) Will the restraint method minimize the danger to the handler? (2) Will the restraint method minimize the danger to the animal? (3) Will the method cause unnecessary pain or fright? (4) Will the restraint method allow the management technique to be completed as necessary? If any one of the question is answered negatively, an alternative method must be chosen.

Whatever the method of restraint selected, it is the manager's responsibility to become proficient at it. Since the majority of people are not farm-raised today, proficiency in restraining animals is not passed from one generation of stockman to the next, as it was in times past. Diligent study of the available restraint methods, a thorough understanding of the animal's anatomy, physiology, and psychology, demonstrations from correctly experienced livestock managers, and then practice on your own are the only ways to acquire the expertise necessary to perform a restraint method and a management technique safely, correctly, quickly, and painlessly as possible.

3. New Words and Phrases

livestock [ˈlaɪvstɑk] *n.* 牲畜；家畜

restraint [rɪˈstrent] *n.* 抑制，克制；约束

psychological [ˌsaɪkəˈlɑdʒɪkl] *adj.* 心理的；心理学的；精神上的

handler [ˈhændlə] *n.* 处理者；管理者；（犬马等的）训练者

exert [ɪɡˈzərt] *vt.* 运用，发挥；施与影响

restriction [rɪˈstrɪkʃən] *n.* 限制；约束；束缚

chemical [ˈkɛmɪkl] *n.* 化学制品，化学药品；*adj.* 化学的

agent [ˈeɪdʒənt] *n.* 代理人，代理商；药剂；特工；*vt.* 由……做中介；由……代理；*adj.* 代理的

physical [ˈfɪzɪkl] *adj.* 物理的；身体的；物质的；根据自然规律的，符合自然法则的；*n.* 体格检查

domesticate [dəˈmɛstɪket] *vt.* 驯养；教化；引进；*vi.* 驯养；使习惯于或喜爱家务和家庭生活

alter [ˈɔːltər] *vt.* 改变，更改；*vi.* 改变；修改

fencelike [ˈfenslaik] *adj.* 像栅栏的，像围栏的，像篱笆的，像围墙的

enclosure [ɪnˈkloʊʒər] *n.* 附件；围墙；围场

welfare [ˈwɛlˈfɛr] *n.* 福利；幸福；福利事业；安宁；*adj.* 福利的；接受社会救济的

minister [ˈmɪnɪstər] *n.* 部长；大臣；牧师；*vi.* 执行牧师职务；辅助或伺候某人

beasts of burden *n.* 役畜；牲口；驮兽

sentiment [ˈsɛntɪmənt] *n.* 感情，情绪；情操；观点；多愁善感

inflict [ɪnˈflɪkt] *vt.* 造成；使遭受（损伤、痛苦等）；给予（打击等）

lose sight of 看不到，看不见；失去与……的联系；忘记；忽略

noxious [ˈnɑkʃəs] *adj.* 有害的；有毒的；败坏道德的；讨厌的

alleviate [əˈliːvieɪt] *vt.* 减轻，缓和

manipulate [məˈnɪpjulet] *vt.* 操纵；操作；巧妙地处理；篡改

alternative [ɔːlˈtɜːrnətɪv] *adj.* 供选择的；选择性的；交替的；*n.* 二中择一；供替代的选择

stockman [ˈstɑkmən] *n.* 畜牧业者；仓库管理员

diligent [ˈdɪlɪdʒənt] *adj.* 勤勉的；用功的，费尽心血的

anatomy [əˈnætəmi] *n.* 解剖；解剖学；剖析；骨骼

expertise [ˈɛkspərˈtiz] *n.* 专门知识；专门技术；专家的意见

4. Notes to the Text A

(1) Restraint can vary from the psychological control that a handler's voice may exert his animals to the complete restriction of activity and total immobilization that chemical agents can provide.

对动物的限制方式可以多种多样，从操作者用声音作用动物进行心理控制，到利用化学药剂对动物的活动和移动进行完全控制。

此句主要学习短语“vary from…to…”，中文含义是“从……到……多种多样”。

(2) Restraint practices became important thousands of years ago when man first domesticated animals for food and fiber and as beasts of burden.

在几千年前，人类一开始驯养动物用于食物、皮毛和役用，控制动物的实践就变得极其重要。

此句“fiber”一词原意是纤维，在此处指动物的皮毛，特别是指毛，如羊毛和兔毛。

(3) Not until many years later was there any concern about the appropriateness of the restraint for the task at hand.

直到多年以后，人们才开始考虑已有的限制措施是否适当。

当“Not until”位于句首时，句子要倒装。其结构为：“Not until+从句/表时间的词+助动词+（主句）主语+谓语+……”。

(4) Today, we realize that it is our responsibility, since we have total control over the animal's life, to be concerned for its welfare, its sensation of pain, and its psychological well-being as they are affected by our production systems and our management techniques.

今天，因为我们完全控制了动物的生活，它们受到我们生产体系和管理技

术的影响，我们要意识到关心动物的福利，关心动物对痛苦的感觉，以及关心动物的心理健康是我们的责任。

句中“since we have total control over the animal's life”是插入语，意思是“因为我们完全控制了动物的生活”。“as they are affected by our production systems and our management techniques”是状语从句，意思是“由于它们受到我们生产体系和管理技术的影响”。

(5) The animal manager's responsibility is to perform his tasks of caring for the animals in the most currently appropriate manner, while inflicting the least amount of pain and causing the least amount of psychological upset (fright).

动物管理者的职责是用最适当的方式，来执行照顾动物的任务，同时使动物遭受最少的痛苦和最小的心理困扰（恐惧）。

(6) Anyone failing to realize this or failing to do everything within his power to alleviate the pain and fright should not be allowed to manipulate the lives of animals.

任何没有意识到这一点的人，或者没有在力所能及的范围内尽力减轻动物的痛苦和恐惧的人，都不应该被允许操控动物的生命。

(7) Diligent study of the available restraint methods, a thorough understanding of the animal's anatomy, physiology, and psychology, demonstrations from correctly experienced livestock managers, and then practice on your own are the only ways to acquire the expertise necessary to perform a restraint method and a management technique safely, correctly, quickly, and painlessly as possible.

勤奋学习可用的管控方法，全面了解动物解剖学、动物生理学、动物心理学，由经验丰富的畜牧管理者正确地演示，然后自己尽可能地熟练掌握这些技术和操作，包括动物的管控方法和安全、正确、快速、无痛管理技术，这是获得必要管控专业知识唯一方法。

5. Exercises

(1) Fill in the blanks by finishing the sentences according to the passage.

①Some degree of control over an animal's movement and activity is required for every technique that the livestock producer ________.

②This ________ altered the natural lifestyle of this animals, and forced man to be responsible for the animal's needs.

③Certain ________ techniques must be performed to achieve those ends. Some of these ________ will cause a degree of pain.

④Pain is a necessary ________ of nature that signals to the animal that something out of the ordinary is happening to its body.

⑤When the animal is ________ and the manager is performing necessary management techniques, there is no ________ from the pain for the animal, no relief from the fright.

⑥Whatever the method of restraint ________, it is the manager's responsibility to become proficient at it.

(2) Answer the following questions according the passage.

①Why is it that every technique the livestock producer performs has some degree of control over animal movements and activities?

②When did the restraint practices of animals became important?

③How are animals affected by our production systems and our management techniques.

④What's questions should you think of when it is your responsibility to select the restraint method to be used in performing a given management technique.

⑤Why should the manager be proficient at the method of restraint selected?

⑥What's the psychological principles of animals underlying physical restraint techniques?

(3) Translation of the following sentences into Chinese.

①With large, potentially dangerous animal, a combination of psychological, physical, and chemical restraint is often employed.

②Managing these animals for human purposes necessitated control of the beasts; there is evidence of early restraint in crude fencelike enclosures.

③However, we should not overemphasized the sentiment of not causing the animal any discomfort and lose sight of the fact that animals are housed, maintained, and ministered to for the production of food, fiber, and pleasure.

④Without this sensation of pain, noxious factors could destroy an animal's body without his knowing it.

⑤Since the majority of people are not farm - raised today, proficiency in restraining animals is not passed from one generation of stockman to the next, as it was in times past.

课文A 动物的控制

动物生产者执行的每一项技术都需要对动物的运动和活动进行某种程度的控制。对动物的控制方式可以多种多样，从操作者用声音对动物进行心理控制，到利用化学药剂对动物的活动和移动进行完全控制。对于大型的，有潜在危险的动物，通常使用心理、物理和化学方法相结合进行控制。

在几千年前，人类一开始驯养动物用于食物、皮毛和役用，控制动物的实践就变得极其重要。这种驯养改变了动物的自然生活方式，迫使人类对动物的所有需要全面负责。为了人类的目的管理这些动物，首先需要控制这些动物；有证据表明，早期人类用类似于围栏的粗糙设施控制动物。直到许多年之后，人们才开始考虑已有的限制措施是否适当。

今天，因为我们完全控制了动物的生活，它们受到我们生产体系和管理技术的影响，我们要意识到关心动物的福利，关心动物对疼痛的感觉，以及关心动物的心理健康都是我们的责任。然而同时，我们也不能过分强调驯养后，没有给动物带来任何不适的情绪，从而忽略了我们将动物舍饲、控制和监视起来生产食物、纤维以及供人类享乐这样一个事实。为了达到驯养目的，人类必须采用某些管理技术，这些技术会不同程度造成动物痛苦。

疼痛是动物发出的一种自救信号，表明身体正在发生某种不正常事情，这是自然界的一种必然现象。如果没有这种痛苦的感觉，有害因素会在动物不知不觉中将其毁灭。动物管理者的职责是用目前最适当的方式，来执行照顾动物的任务，同时使动物遭受最少的痛苦和最小的心理困扰（恐惧）。当动物受到控制，管理者正执行必要的管理技术时，动物就无法摆脱痛苦，恐惧也无法得到缓解。任何意识不到这一点的人，或者没有在力所能及的范围内尽力减轻动物痛苦和恐惧的人，都不应该被允许操控动物的生命。

当你负责选择控制动物的方法和管理技术时，你必须问自己以下问题：(1) 这种控制方法对操作者的危险是最小的吗？(2) 这种控制方法对动物的危险是最小的吗？(3) 这种方法会引起动物不必要的痛苦或恐惧吗？(4) 这种控制能保证完成必要的管理技术吗？如果有一个问题的答案是否定的，就必须选择其他方法。

无论选择何种控制方法，管理者都有责任精通这种方法。由于今天大多数人都不是在农场长大的，控制动物的专业技术不再像过去那样，由一代畜牧人传给下一代。因此勤奋学习可用的管控方法，全面了解动物解剖学、动物生理学、动物心理学，由经验丰富的畜牧管理者正确地示范，然后自己尽可能地熟练掌握这些技术和操作，包括动物的管控方法和安全、正确、快速、无痛管理技术，这是获得必要管控专业知识唯一方法。

Part Two Extensive Reading

1. Text B Technology of Animal Restraint

Types of restraint

There are five categories of restraint: (1) psychological, (2) sensory diminishment, (3) use of confining chutes, alleys, and barriers, (4) use of tools and physical force, and (5) chemical sedation or immobilization. Each of these has its advantages and disadvantages, depending upon the species of livestock involved and the management technique to be performed.

Psychological restraint depends upon the manager having a thorough working knowledge of the behavior patters of the species to be restrained. With this knowledge, the manager can take steps to either make use of or offset the animal's natural behavioral tendencies. For example, when working a group of sheep, the experienced shepherd appreciates the futility of blindly trying to rush them through an opening and, instead, calmly "work" them until one of the flock starts through. At that point, the others will follow with a minimum coercion. Another example, this time using a combination of psychological restraint and a tool, would be the proper use of a hog snare to restrain a pig. It is the natural tendency of the pig to pull backward against the pull of the snare. The knowledgeable manager would allow the pig to pull backward until it has positioned itself into a corner, thus maneuvering itself into being restrained from both side.

Sometimes the human voice can actually be used as a restraint tool depending on the previous conditioning of the animals. Authority (or the lack of it), confidence (or fear), a coincidence soothing (or exciting) effect can all be transmitted in a voice. Animals can readily perceive this and they do respond to it.

More than likely, the animal is actually responding to the combination of voice and mannerisms of the manager. The manager must move confidently and quickly, but not with the false bravado shouting, frantic arm waving, and jumping about. Self-confidence in one's ability to get the job done comes through naturally, and if it's there, the animal will respond accordingly. The only things that you can do develop this self-confidence are to study the animal's behavior, anatomy, and physiology, watch the techniques being performed properly by others, practice on your own, and then believe in your ability.

Sensory diminishment as a way of restraining animals usually involves blindfolding. Under certain conditions, such as when animals must be maintained in ultra noisy surroundings that continually excite them, plugging the ears with cotton often has a quieting effect.

Blindfolding sometimes works, but be cautioned and aware that just because a horse is blindfolded does not mean that it will blindly enter a trailer that it has resisted for hours! Neither will the horse suddenly allow you to clip its ears with bussing electric clipper just because you blindfold it (Inserting large cotton balls into the ear, however, may just be the extra edge you need to make the horse submit to clipping). Blindfolding is probably best utilized as an adjunct restraint to assist you in quieting and controlling domestic animals that are resisting the primary restraint in too violent a manner.

The use of confining alleys, chutes, and barriers is one of the most common ways to restrain domestic livestock, especially cattle and sheep. However, before the manager makes the decision to use the method as his primary restraint, he must once again call upon his knowledge of animals behavior, anatomy, and physiology, and upon his own common sense. For example: 18 kg lambs do not need a squeeze chute for vaccinating; hogs are more easily worked with a snare for blood testing than with a head gate; horse would "blow up" if squeezed in a squeeze chute or if their heads were "taken away" from them in ahead gate, but cattle should be placed in a head gate with a nose bar for dehorning. There is no substitute for knowledge, experience, and common sense when making these decisions.

2. Notes to the Text B

(1) The knowledgeable manager would allow the pig to pull backward until it has positioned itself into a corner, thus maneuvering itself into being restrained from both side.

聪明的操作者控制猪时，会利用猪的天性让猪往后退，直到猪自己退到一个角落，使其从两边都被控制。

句中"until"引导时间状语从句，表示"直到……才……"。

(2) The manager must move confidently and quickly, but not with the false bravado shouting, frantic arm waving, and jumping about. Self-confidence in one's ability to get the job done comes through naturally, and if it's there, the animal will respond accordingly.

管理者操作必须自信而迅速，但不能虚假或虚张声势地叫喊，也不能疯狂地手舞足蹈和上蹿下跳。只有一个人在工作中充满自信，才能更好地控制动物，动物也会做出相应的反应。

此句“and if it's there, the animal will respond accordingly”是由“and”引导的一个并列句，两个分句间有“and”等并列连词连接，这时，后一分句常用人称代词或“it”,“that”代指前一分句中的某一名词或整个句子的内容，构成并列句式结构。后一分句中，“if it's there”也是由“if”引导的状语从句。

（3）Blindfolding is probably best utilized as an adjunct restraint to assist you in quieting and controlling domestic animals that are resisting the primary restraint in too violent a manner.

蒙眼最好是作为辅助控制措施，蒙眼能使动物安静，有助于控制强力反抗的动物。“that are resisting the primary restraint in too violent a manner”是“that”引导的定语从句，修饰“animals”。

3. Answer Questions

（1）How many categories of restraint animals do they have?

（2）How to use human voice as a tool of restraint animals?

（3）What should we pay attention to when we restrain animals with a blindfold?

（4）What is the most common way to restrain domestic livestock?

课文B　动物的保定技术

保定的方法

保定是对动物的控制，可以分为五类：(1) 心理控制；(2) 降低感知；(3) 使用封闭的围栏、通道和障碍物；(4) 使用工具和外力；(5) 化学镇静或固定作用。每一种控制方法都有其优缺点，这取决于所要控制的动物以及对所用控制技术的具体操作过程。

心理控制取决于操作者对被控制对象动物的行为模式有丰富工作知识和深入了解。有了这方面的知识，操作者可以利用动物的自然行为倾向或抵消、打破动物的自然行为倾向。例如，当控制一群羊时，经验丰富的牧羊人不会盲目地驱赶整个羊群通过一个通道，他们知道这样做是徒劳无益的。相反，他会平静地使其中一只羊开始通过，羊群中其他羊就会尾随而过，这样就不用费力进行驱赶羊群。另一个例子是使用心理控制与工具控制相结合的方法，是正确使用猪诱捕器控制猪。猪的天性是后退，它会向诱捕器方向使劲后退。聪明的操作者控制猪时，会利用猪的天性让猪往后退，直到猪自己退到一个角落，使其从两边都被控制。

如果动物事前经过训练，人类的声音实际上也可以被用作一种控制工具。管理者对动物有无权威、管理者对动物是自信或恐惧、管理者对动物的安慰或刺激都可以通过自己的声音传递给动物，动物能够很容易地感知到这一点，并且做出反应。

实际上更有可能的是动物对管理者声音和举止做出综合反应。管理者操作必须自信且迅速，但不能用虚假或虚张声势的叫喊，也不能疯狂地手舞足蹈和上蹿下跳。只有一个人在工作中充满自信，才能更好地控制动物，动物也会做出相应反应。你要提高在控制动物过程的自信，唯一的方法就是学习动物的行为学、解剖学和生理学，观察别人正确运用的技巧，自己不断练习和提高，同时还要充分相信自己的能力。

蒙住眼睛是在控制动物时的一种降低感知的常用方法。在某些条件下，例如当动物在极端嘈杂的环境中生存时，长时间被噪音刺激，可以用棉花塞住耳朵，这样通常会产生安静的效果。

蒙眼有时是有效的，但是要注意，仅仅因为一匹马被蒙眼并不意味着它会盲目地进入一辆它抵抗了几个小时的拖车！这匹马也不会因为你蒙上眼睛就突然允许你用公共电动剪剪耳朵（然而，将大棉球插入马的耳朵，对马进行剪耳，马就较容易服从操作）。蒙眼最好是作为辅助控制措施，蒙眼能使动物安静，有助于控制强力反抗的动物。

使用封闭的通道、围栏和障碍物是控制家畜的常见方法之一，特别是控制牛羊时最常用。然而，在操作者决定使用某种控制方法作为主要方法之前，必须再次回顾其动物行为学、解剖学和生理学的相关知识以及他自己平时掌握常识。例如，18kg 重的羔羊接种时不需要用控制车；猪血液测试时，诱捕工具比头保定门更容易操作；如果马被挤压入一个挤压通道，或者用卡脖门卡住脖颈，马就会“惊爆”，但牛去角时，应该使用一个带牛鼻钳的卡脖门进行控制。在做出如何对动物进行控制的决定时，经验和常识是无可替代的。

扫码进行拓展学习

Lesson 11

Ecology and Environment

Part One Intensive Reading

1. Learning Objective

After learning this lesson, you should understand the following:

(1) Define the terms of ecology and environment.

(2) Describe the important role of environment and ecology in animal production.

(3) Give an overview of which factors the production environment includes.

(4) Explain why domesticated animals best suited to a particular environment survived.

(5) Give an overview of why it is becoming increasingly difficult to define environment.

(6) Describe why the keepers of herds and flocks were more and more concerned with the effect of environment on animals today.

(7) Describe what animals environmental control involves.

2. Text A Livestock Environment and Ecology

Among animals, environmental control involves space requirements, light, air temperature, relative humidity, air velocity, wet bedding, ammonia buildup, dust, odors, and manure disposal, along with proper feed and water. Control or modification of these factors offers possibilities for improving animal performance. Although their there is still much to be learned about environmental control, the gap between awareness and application is becoming smaller. Researched on animal environment has lagged, primarily because it requires a melding of several disciplines—nutrition, physiology, genetics, engineering, and climatology. Those engaged in such studies are known as ecologists. An animal is the result of two forces, heredity and environment. Heredity has already made its contribution at the time of fertilization, but environment works ceaselessly away until death. Since most animal traits are only 30% to 50% heritable, the expression of the rest (more than 50%) depends on the quality of all components of the environment. Thus, it is very important that the keeper of herds and flocks have enlightened knowledge of, and apply expert management to, animal environment.

Environment may be defined as all the conditions, circumstances, and influences surrounding and affecting the growth, development, and production of animals. The most important influences in the environment are the feed and quarters (space and shelter).

The branch of science concerned with the relation of living things to their environment and to each other is known as ecology. Ecology is also defined as the scientific study of the interactions of organisms with the environment that determine their distribution and abundance. The environment of an organism consists of all those factors and phenomena that can influence it, whether those factors be physical and chemical (abiotic) or other organism (biotic).

Animal ecology as a central discipline of biology overlaps principally with four other areas of study. They are physiology, ethology, evolution and genetics.

Livestock ecology is the scientific study of the interactions of domestic animals with their production environment and their effects on numbers, density and distribution of livestock. The production environment includes all external factors and phenomena that affect the behavior of livestock system. These can be of an abiotic, biotic, economic or socio-cultural nature.

Through the years, the domesticated animals best suited to a particular environment survived, and those that were poorly adapted either moved to a more favorable environment or perished. During the past two centuries, livestock producers have made great strides in the selection and propagation of animals suited to a particular environment, and during the past 50 years they have made progress in modifying the environment for the benefit of their animals and themselves.

It is becoming increasingly difficult to define environment, because scientists continue to discover important new environmental factors. Primitive people recognized the sun and the fire provided both heat and light, that body heat could be conserved by draping the body with animal skins, and that trees and caves protection from the weather. Today, it is recognized that these, along with a host of other environmental factors affect animals and people.

The keepers of herds and flocks were little concerned with the effect of environment on animals so long as grazed on pastures or ranges. But rising feed, land, and labor costs along with concentration of animals into smaller spaces, changed all this. Today, most layers and broilers are on litter floors. Turkeys are shifting rapidly from range to confinement. Water is important for ducks, but even with ducks the trend is toward higher population densities and more confinement. Many swine are raised partially or totally in confinement; and confinement production is increasing with beef cattle, dairy cattle, and sheep.

Among animals, environmental control involves space requirements, light, air temperature, relative humidity, air velocity, wet bedding, ammonia buildup, dust, odors, and manure disposal, along with proper feed and water. Control or modification of these factors offers possibilities for improving animal performance. Although there is still much to be learned about environmental control, the gap between awareness and application is becoming smaller. Researched on animals environment has lagged, primarily because it requires a melding of several disciplines—nutrition, physiology, genetics, engineering, and climatology. Those engaged in such studies are known as ecologists.

3. New Words and Phrases

ecology [ɪˈkɑlədʒi] *n.* 生态学；生态

heredity [həˈredəti] *n.* 遗传；遗传特征

environment [ɪnˈvaɪrənmənt] *n.* 环境，外界；周围，围绕；工作平台；（运行）环境

fertilization [ˌfɜrtəlɪˈzeɪʃən] *n.* 施肥；受精，受精过程，受精行为；受孕；受胎

ceaselessly [ˈsiːslisli] *adv.* 不停地，持续地；不住

trait [treɪt] *n.* 特点，特性；少许；性状

expression [ɪkˈsprɛʃən] *n.* 表现，表示，表达；表情，脸色，态度，腔调，声调

enlighten [ɪnˈlaɪtn] *v.* 启发；开导；教导

development [dɪˈvɛləpmənt] *n.* 发展，进化；被发展的状态；新生事物，新产品；发育

quarters [ˈkwɔtərz] *n.* 住处，岗位

branch [bræntʃ] *n.* 分部；部门；分店；分支；树枝 *v.* 分岔；分支

concern [kənˈsɜːrn] *vt.* 涉及，关系到；使担心 *n.* 关系；关心；关心的事；

忧虑

determine [dɪˈtɝmɪn] *v.* （使）下决心，（使）做出决定 *vt.* 决定，确定；判定，判决；限定 *vi.* 确定；决定；判决，终止

distribution [ˈdɪstriˈbiuːʃn] *n.* 分布；分发；分配；散布；销售量

phenomena [fəˈnɒmɪnə] *n.* 现象，phenomenon 的复数形式

abiotic [ˌeɪbəɪˈɒtɪk] *adj.* 无生命的，非生物的

biotic [baɪˈɑːtɪk] *adj.* 有关生命的；生物的

discipline [ˈdɪsəplɪn] *n.* 纪律；训练；学科 *vt.* 训练；惩罚

principally [ˈprɪnsəpli] *adv.* 主要地；大部分

ethology [iːˈθɒlədʒɪ] *n.* 动物行动学，道德体系学，个体生态学

interaction [ˌɪntəˈrækʃən] *n.* 一起活动；合作；互相影响；互动

domestic [dəˈmɛstɪk] *adj.* 家庭的；国内的；驯养的

perished [ˈpɛrɪʃt] *adj.* 感觉很冷的；脆裂的 *v.* 灭亡（perish 的过去分词）；枯萎

propagation [ˌprɑpəˈgeʃən] *n.* 传播；繁殖；增殖

modify [ˈmɑdɪfaɪ] *vt.* 修改，修饰；更改 *vi.* 修改

primitive [ˈprɪmətɪv] *adj.* 原始的；简陋的 *n.* 文艺复兴前的艺术家；原始人

ranges [reɪndʒ] *n.* 范围；围栏牧场；射程；山脉；排；一系列；闲逛；炉灶 *v.* 排列；使……站在某一方；延伸；漫游

humidity [hjuˈmɪdəti] *n.* 湿度；湿气

ammonia [əˈmoʊniə] *n.* 氨

buildup [ˈbɪldʌp] *n.* 集结；增长；树立名誉

odor [ˈoʊdər] *n.* 气味；名声；气息

manure [məˈnʊr] *vt.* 施肥于；耕种 *n.* 肥料；粪肥

improve [ɪmˈpruv] *vt.* 提高（土地、地产）的价值；利用（机会）；改善，改良 *vi.* 变得更好；改进，改善

engage [ɪnˈgedʒ] *vt.* 吸引，占用；使参加；雇佣；使订婚；预订 *vi.* 从事；答应，保证；交战；啮合

legislator [ˈledʒɪsleɪtər] *n.* 立法委员；立法者

4. Notes to the Text A

（1）Heredity has already made its contribution at the time of fertilization, but environment works ceaselessly away until death.

句中“but”引导的并列句，表示转折。句中“contribution”和“work”

含义都是“作用”。

(2) Since most animal traits are only 30 to 50% heritable, the expression of the rest (more than 50%) depends on the quality of all components of the environment.

此句中“most animal traits”是指“动物的多数性状”，而并非“多数动物的形状”。

(3) The most important influences in the environment are the feed and quarters (space and shelter).

此句中“quarters”一词原意是“住处和岗位”，此处是指“动物的圈舍”。

(4) The branch of science concerned with the relation of living things to their environment and to each other is known as ecology.

此句中“concerned with”是指有关的，“the relation of living things to their environment and to each other”是“relation…to…to…”结构，表示相互之间的关系，此处指“生物与环境以及生物之间的相互关系”。

(5) Through the years, the domesticated animals best suited to a particular environment survived, and those that were poorly adapted either moved to a more favorable environment or perished.

此句是一个复杂句型，“and”引导的并列句中，在第一个分句“the domesticated animals best suited to a particular environment survived”中“best suited to a particular environment”是一个省略了“that”的定语从句，在第二个分句“and those that were poorly adapted either moved to a more favorable environment or perished”中，又由“that”引导的定语从句修饰“those”，“those”是指“the domesticated animals”。第二个分句使用了“either…or…”句型，此句型可以用来连接并列的主语、谓语或宾语，本句连接并列的谓语。

(6) Ecology is also defined as the scientific study of the interactions of organisms with the environment that determine their distribution and abundance.

此句中“the interactions of…with…”意思是“……与……的相互作用”，“that determine their distribution and abundance”是由“that”引导的定语从句，修饰“environment”，意思是“可以决定分布和数量的环境”。

(7) Turkeys are shifting rapidly from range to confinement.

此句“range”原意为“一定范围”，这里指“放牧”，“confinement”原意为“限制”，这里指“圈养”。

(8) More extensive methods of farming, *e.g.* free range, can also raise welfare concerns such as the mulesing of sheep, predation of stock by wild animals,

and biosecurity.

此句“mulesing”为澳大利亚给羊做的一种手术。

5. Exercises

(1) Fill in the blanks by finishing the sentences according to the passage

①An animal is the result of two forces, heredity and environment. Heredity has already made its contribution at the time of ________ , but environment works ceaselessly away until ________ .

②The most important influences in the ________ are the feed and quarters.

③The branch of science concerned with the relation of living things to their environment and to each other is known as ________ .

④The environment of an organism consists of all those factors and phenomena that can influence it, whether those factors be ________ and ________ (abiotic) or other ________ (biotic) .

⑤Animals ________ as a central discipline of biology overlaps principally with four other areas of study.

⑥Livestock ________ is the scientific study of the interactions of domestic animals with their production environment and their effects on numbers, density and distribution of livestock.

⑦The keepers of ________ and ________ were little concerned with the effect of environment on animals so long as grazed on pastures or ranges.

⑧Water is important for ducks, but even with ducks the trend is toward higher population ________ and more ________ .

(2) Answer the following questions according to the passage

①What is the difference between ecology and environment?

②What is the important role of environment and ecology in animal production?

③What does the animal production environment include?

④Why can domesticated animals best suited to a particular environment survive?

⑤Whyis environment becoming increasingly difficult to define?

⑥Why is the keepers of herds and flocks more and more concerned with the effect of environment on animals today?

⑦How to control the animals environment?

(3) Translation of the following sentences into Chinese

①Since most animal's traits are only 30% to 50% heritable, the expression of the rest (more than 50%) depends on the quality of all components of the environment. Today, most layers and broilers are on litter floors.

②Ecology is also defined as the scientific study of the interactions of organisms with the environment that determine their distribution and abundance.

③Livestock ecology is the scientific study of the interactions of domestic animals with their production environment and their effects on numbers, density and distribution of livestock.

④But rising feed, land, and labor costs along with concentration of animals into smaller spaces, changed all this.

⑤Researched on animal environment has lagged, primarily because it requires a melding of several disciplines—nutrition, physiology, genetics, engineering, and climatology.

课文A 家畜的环境与生态

在动物中，环境控制包括空间需求、光线、气温、相对湿度、气流速度、湿卧床、氨积聚、灰尘、气味、粪便处理，以及适当的饲料和水。控制或改变这些因素，可改善动物性能。虽然生态研究人员在环境控制方面还有很多需要学习的地方，但是知识和应用之间的差距正在缩小。对动物环境的研究一直滞后，主要是因为它需要几个学科的融合——营养学、生理学、遗传学、工程学和气候学。从事这类研究的人被称为生态学家。

动物是遗传和环境两种力量的产物。遗传在其受孕时作用已经确定，但是环境的作用持续终生。由于动物的大多数性状只有30%~50%遗传力，其性状表达超过50%取决于环境因素。因此，动物的饲养者必须对动物环境有明确认识，并对动物环境进行专业管理，这是很重要的。

环境可定义为动物周围影响其生长、发育和生产的所有条件、境况和影响因素。环境中最大的影响因素是饲料和圈舍（空间和棚舍）。

关于生物本身、生物环境及其各因素之间关系的科学分支被称为生态学。生态学是研究决定生物分布和丰度的环境与生物之间相互作用的科学。一个生物的环境由所有那些可以影响它的因素和现象组成，无论这些因素是物理的、化学的还是其他生物的。

动物生态学作为生物学的一个核心学科，主要与其他四个研究领域重叠。它们是生理学、行为学、进化论和遗传学。

家畜生态学是研究家畜与其生产环境的相互作用及其对牲畜数量、密度和分布影响的科学。生产环境包括影响牲畜系统行为的所有外部因素和现象，它们可以是非生物的、生物的、经济的或社会文化的。

随着时间的推移，最适合特定环境的驯养动物得以存活，而那些适应能力不佳的动物要么转移到了更有利的环境中，要么就灭绝了。在过去两个世纪中，牲畜生产者在选择和繁殖适合特定环境的动物方面取得了巨大进展，而在过去50年中，为了动物和自身利益，其在改善环境方面取得了进展。

由于科学家不断发现新的重要环境因素，环境定义变得越来越困难。原始人类认识到太阳和火都能提供热和光，用动物毛皮包裹身体，可以保存体热，树木和洞穴可以保护自己不受天气影响。今天，人们认识到，这些环境因素以及许多其他环境因素都影响动物和人类。

只要在草场或牧场上放牧牛群和羊群，饲养者就极少关心环境对动物的影响。但是，随着饲料、土地和劳动力成本的不断上升，动物被聚集到更小空间集约饲养而使情况彻底改变。如今，大多数的蛋鸡和肉鸡都在垫草地板上。火鸡正在迅速地从放牧转移到圈养。水对鸭子很重要，即使如此，养鸭的趋势也是朝向更高的种群密度和更多的人为控制。多数猪是部分或全部时间进行圈养的；而肉牛、奶牛和绵羊的圈养生产方式也在增加。

动物的环境控制涉及空间要求、光线、空气温度、相对湿度、空气流速、卧床湿度、氨浓度、灰尘、气味和粪便处理，以及适当的饲料和水。这些因素的控制和改变为提高动物性能提供了可能性。尽管在环境控制方面仍有许多东西需要学习，但意识与应用之间的差距正在缩小。对动物环境的研究滞后，主要是因为它需要几个学科的融合——营养、生理学、遗传学、工程学和气候学。从事这类研究的人被称为生态学家。

Part Two Extensive Reading

1. Text B Effect of Environmental Factors on Animals

Preventing disorders by merely cutting off the tails of pigs to alleviate tail biting, debeaking poultry to prevent cannibalism, and using choke collars on horses to inhibit cribbing, is not unlike trying to control malaria fever in humans by the use of drugs without getting rid of mosquitoes. Rather, we need to recognize these disorders for what they are warning signals that condition are not right. Correcting the cause of the disorder is the best solution. Unfortunately, this is not usually the easiest. Correcting the cause may involve trying to emulate the natural conditions of the species, such as altering space per animal and group size, providing training and experience at opportune times, promoting exercise, and gradually changing rations. Over the long pull, selection provides a major answer to correcting confinement and other behavioral problems; we need to breed animals adapted to people-made environments. The following factors are of special importance in any discussion of animal's environment: feed, water, weather, facilities, health, and stress.

Feed

Animals may be affected by either (1) too little or too much feed, (2) rations that are too low in one or more nutrients, (3) an imbalance between certain nutrients, or (4) objection to the physical form of the ration, for example, it may be ground too finely.

Forced production (such as growth, milk products, and racing 2-years-old horse) and the feeding of forages and grains which are often produced on leached and depleted soils have created many problems in nutrition. These conditions have been further aggravated through the increased confinement of animals, many animals being confined to stalls or lots all or large part of the year. Under these unnatural conditions, nutritional diseases and ailments have become increasingly common.

Also, nutritional reproductive failures plague livestock operations. Generally speaking, energy supply tends to be more limiting than protein in reproduction. The level and kind of feed before and after parturition determine how many females will heat and conceive. After giving birth, feed requirements increase tremendously because of milk production; hence, a female suckling young needs approximately 50% greater feed allowance than during the pregnancy period. Otherwise, she will suffer serious loss in weight, and she may fail to come in heat and conceive. This basic fact, along with other pertinent finding, was confirmed by researchers at the Montana Agricultural Experiment Stations. Based on 12 years research at the Havre and Miles City Stations, they concluded that beef cattle size and milk production should be tailored to fit the environment. Big size and more milk are not better unless the range forage supply is better. The best size cow is one that fits the range conditions. Small cows do on poor range because they can usually get 100% of their daily feed requirement for maintenance and milk production, whereas big cows on a poor range are borderline hungry all the time. Also, cows that give a lot of milk must have a good range; otherwise, they are stressed by lack of feed; and their fertility rate and calf crops drop. So, cow size and milk production should match their environment.

The next question is whether a breeding program can make maximum progress under conditions of suboptimal nutrition (such as is often found under some farm and range conditions). One school of thought is that selection for such factors as body form and growth rate in animals can be most effective only under nutritive conditions promoting the near maximum development of those characters of which the animal is capable. The other school of thought is that genetic differences affecting usefulness under suboptimal conditions will be expressed under such suboptimal conditions, and that differences observed under forced conditions may not be correlated with real utility under less favorable conditions. Those favoring the latter thinking argue, therefore, that the production and selection of breeding animals for the range should be under typical range conditions and that the animals should not be highly fitted in a box stall.

The results of a 10-year experiment conducted by the senior author and his colleagues at Washington State University, designed to study the effect of plane of nutrition on meat animal improvement, support the contention that selection of breeding animals should be carried on under the same environmental conditions as those under which commercial animals are produced.

2. Notes to the Text B

(1) Preventing disorders by merely cutting off the tails of pigs to alleviate tail biting, debeaking poultry to prevent cannibalism, and using choke collars on horses to inhibit cribbing, is not unlike trying to control malaria fever in humans by the use of drugs without getting rid of mosquitoes.

此句“cutting off the tails of pigs”是断尾，在集约化养猪生产中为了防止猪相互咬尾巴，一般在仔猪出生后将其尾巴剪掉。“debeaking”是在蛋鸡集约化生产中为了防止其相互间打斗和啄食，雏鸡出壳后用专用断喙器将喙切掉一部分。这句话在批评集约化生产过程中“头痛医头脚痛医脚”的做法。

(2) Animals may be affected by either ① too little or too much feed, ②rations that are too low in one or more nutrients, ③an imbalance between certain nutrients, or ④ objection to the physical form of the ration, for example, it may be ground too finely.

此句“rations”是指“日粮”，“objection”是指“缺陷”。

(3) Forced production (such as growth, milk products, and racing 2-year-old horse) and the feeding of forages and grains which are often produced on leached

and depleted soils have created many problems in nutrition.

此句中“which are often produced on leached and depleted soils”是定语从句，“which”指“forages and grains”。

（4） Also, nutritional reproductive failures plague livestock operations.

此句中“nutritional reproductive failures”意思是“营养性繁殖障碍”。

（5） Based on 12 years research at the Havre and Miles City Stations, they concluded that beef cattle size and milk production should be tailored to fit the environment.

此句“Havre and Miles”是蒙大拿州的两个城市，“tailor”意思是“使合适”。

（6） One school of thought is that selection for such factors as body form and growth rate in animals can be most effective only under nutritive conditions promoting the near maximum development of those characters of which the animal is capable.

此句为复杂句，首先是“that”引导的宾语从句，从句的主语是“selection for such factors as body form and growth rate in animals”。“under nutritive conditions”是介词短语做状语，“promoting the near maximum development of those characters”分词短语做定语修饰“nutritive conditions”。“of which the animal is capable”定语从句修饰“characters”。

3. Answer Questions

（1） How to prevent the disorders of pigs and poultry?

（2） Which factors are of special importance in any discussion of animal environment?

（3） What may animals be affected by?

（4） Can a breeding program make maximum progress under conditions of suboptimal nutrition?

（5） What are the results of a 10-year experiment conducted by the senior author and his colleagues at Washington State University?

课文 B　环境因素对动物的影响

通过断尾来预防猪咬尾疾病，通过给家禽断喙预防相互啄食，以及通过在马身上用索套项圈来抑制躺卧来预防疾病，这和不消灭蚊子，仅利用药物来试图控制人类的疟疾没什么不同。相反，我们需要认识到在哪些不正确的条件下这些疾病会发出警示信号。纠正病因是最好的解决办法，不幸的是，这通常不是最容易的事情。纠正原因可能需要努力模仿物种的自然生存条件，例如改变每个动物生存空间和群体大小，适时进行培训和演示，促进锻炼，以及逐步改变日粮。如何纠正圈养和一些行为问题，经过长时间的努力，得出一个主要答案就是进行选择；我们需要通过育种，培育适应人工环境的动物。在讨论动物与环境时，下列因素特别重要：饲料、水、天气、设施、健康和压力。

饲料

饲料可能影响动物的因素包括（1）饲料过多或过少；（2）日粮中缺少一种或多种营养物质；（3）某些营养之间的不平衡；（4）日粮的物理形式缺陷，例如，饲料磨得过细。

强制生产（如促长、促乳和2岁的赛马）和饲喂产于贫瘠土壤的饲料和谷物，通常造成了许多营养问题。由于对动物的限制增加，使其生存条件进一步恶化，许多动物全年或大部分时间都被限制在畜栏或畜圈中。在这种非自然的条件下，营养性疾病变得越来越普遍。

营养不良繁殖障碍也困扰着畜牧企业。一般来说，能量对繁殖的影响往往比蛋白质更大。分娩前后的饲料营养水平和种类决定畜群母畜的发情和怀孕数量。母畜分娩后，由于产乳的缘故，对饲料的需求大大增加；因此，一个带仔哺乳母畜的饲料需要定额要高出其怀孕期间的50%。否则，其体重将会严重下降，而且出现不发情，难以怀孕的情况。这一基本事实，以及其他相关的发现，得到了蒙大拿农业试验站研究人员的证实。根据阿弗尔城市和迈尔斯市实验站12年的研究，他们得出结论，肉牛的大小和牛乳的生产应该根据环境进行调整。除非牧场的饲料供应较好，牛体大、产牛乳多并不好。最适合牧场条件的奶牛大小是最好的。在较差的牧场，体型小的奶牛通常就可以100%满足维持和产乳的需要，而体型大的奶牛总是处于饥饿边缘。此外，奶牛生产大量牛乳必须有良好的放牧场；否则，母牛会受到缺乏饲料的压力；它们的繁殖率和小牛产量都会下降。因此，奶牛的体型和产乳量应该与环境相匹配。

下一个问题是一个育种计划能否在亚营养条件下取得最大的进展（如在某些农场和牧场条件下）。一种学术观点认为，只有在营养能最大限度促进动物性状发育的条件下，选择动物的体形和生长速度等因素才最有效。另一种学术观点认为，在不理想的条件下，影响遗传差异的效用将被表达出来，而在强制条件下观察到的差异可能与在不利条件下的实际效用无关。赞成后者的人认为，饲养动物的生产和选择应该在典型的牧场条件下，而不应该在高度适宜的动物圈箱中。

有位资深作者和他的同事在华盛顿州立大学进行了一项为期10年的实验，旨在研究营养水平对肉类动物改良的影响，这项实验的结果支持了这样一种观点，即选育动物应该在与动物商业生产相同的环境条件下进行。

扫码进行拓展学习

Lesson 12
Welfare

Part One Intensive Reading

1. Learning Objective

After learning this lesson, you should understand the following:

(1) Give an overview of the standards of "good" animal welfare.

(2) Give an accurate accounting of respect for animal welfare.

(3) Describe the two forms of criticism of the concept of animal welfare.

(4) Identify and place in context the role of animal welfare in farm animals.

(5) Describe the predominant view of modern neuroscientists on consciousness.

(6) Describe the major concern for the welfare of farm animals.

(7) Give an accounting of production parameters which sometimes impinge on the farm animals' welfare.

2. Text A Animal Welfare

Animal welfare is the well-being ofanimals. The standards of "good" animal welfare vary considerably between different contexts. These standards are under constant review and are debated, created and revised by animal welfare groups, legislators and academics worldwide. Animal welfare science uses various measures, such as longevity, disease, immunosuppression, behavior, physiology, and reproduction, although there is debate about which of these indicators provide the best information.

In respect for animal welfare is often based on the belief that non-human animals are sentient and that consideration should be given to their well-being or suffering, especially when they are under the care of humans. These concerns can include how animals are slaughtered for food, how they are used in scientific research, how they are kept (as pets, in zoos, farms, circuses, etc.), and how human activities affect the welfare and survival of wild species.

There are two forms of criticism of the concept of animal welfare, coming from diametrically opposite positions. One view, held by some thinkers in history, holds that humans have no duties of any kind to animals. The other view is based on the animal rights position that animals should not be regarded as property and any use of animals by humans is unacceptable. Accordingly, some animal rights proponents argue that the perception of better animal welfare facilitates continued and increased exploitation of animals. Some authorities therefore treat animal welfare and animal rights as two opposing positions. Others see animal welfare gains as incremental steps towards animal rights.

The predominant view of modern neuroscientists, notwithstanding philosophical problems with the definition of consciousness even in humans, is that consciousness exists in nonhuman animals. However, some still maintain that consciousness is a philosophical question that may never be scientifically resolved.

A major concern for the welfare of farm animals is factory farming in which large numbers of animals are reared in confinement at high stocking densities. Issues include the limited opportunities for natural behaviors, for example, in battery cages, veal and gestation crates, and routine invasive procedures such as beak trimming, castration, and ear notching, instead producing abnormal behaviors such as tail - biting, cannibalism, and feather pecking. More extensive methods of farming, *e. g.* free range, can also raise welfare concerns such as the mulesing of sheep, predation of stock by wild animals, and biosecurity.

Farm animals are artificially selected for production parameters which sometimes impinge on the animals' welfare. For example, broiler chickens are bred to be very large to produce the greatest quantity of meat per animal. Broilers bred for fast growth have a high incidence of leg deformities because the large breast muscles cause distortions of the developing legs and pelvis, and the birds cannot support their increased body weight. As a consequence, they frequently become lame or suffer from broken legs. The increased body weight also puts a strain on their hearts and lungs, and ascites often develops. In the UK alone, up to 20 million broilers each year die from the stress of catching and transport before reaching the slaughterhouse.

Another concern about the welfare of farm animals is the method of slaughter, especially ritual slaughter. While the killing of animals need not necessarily involve suffering, the general public considers that killing an animal reduces its welfare. This leads to further concerns about premature slaughtering such as chick culling by the laying hen industry, in which males are slaughtered immediately after hatching because they are superfluous; this policy occurs in other farm animal industries such as the production of goat and cattle milk, raising the same concerns.

3. New Words and Phrases

longevity [lɔn'dʒɛvəti] *n.* 长寿，长命；寿命

immunosuppression [ˌɪmjunosə'prɛʃən] *n.* 免疫抑制

well-being [ˌwɛl'biɪŋ] *n.* 幸福；生活安宁；福利

suffering ['sʌfərɪŋ] *n.* 受难；苦楚 *adj.* 受苦的；患病的 *v.* 受苦；蒙受（suffer 的 ing 形式）

diametrically [ˌdaɪə'mɛtrɪkli] *adv.* 完全地；作为直径地；直接地；正好相反地

proponent [prə'poʊnənt] *n.* 提倡者；支持者

exploitation [ˌɛksplɔɪ'teʃən] *n.* 开发；开采；剥削；利用

incremental [ɪnkrə'mɑntl] *adj.* 增加的，增值的

neuroscientists [ˌnjʊərəʊ'saɪəntɪst] *n.* 神经系统科学家

notwithstanding [ˌnɑtwɪθ'stændɪŋ] *prep.* 尽管；虽然 *adv.* 尽管如此，仍然；还是 *conj.* 虽然，尽管

consciousness ['kɑnʃəsnəs] *n.* 意识；知觉；自觉；觉悟

philosophical [ˌfɪlə'sɑfɪkl] *adj.* 哲学上的，哲学（家）的；冷静的，沉着的；明达的；达观的

battery ['bætəri] *n.* 电池；一系列；炮兵连；排炮；[律] 殴打

invasive [ɪn'vesɪv] *adj.* 侵略性的，侵害的；攻击性的；扩散性的，蔓延性的

cannibalism ['kænəblˌɪzəm] *n.* 吃人肉的习性；同类相食

impinge [ɪm'pɪndʒ] *vi.* 撞击；侵犯；对……有影响 *vt.* 撞击，打击

deformity [dɪ'fɔrməti] *n.* 畸形；残废

distortion [dɪs'tɔrʃən] *n.* 扭曲，变形；失真，畸变；扭转

pelvis ['pɛlvɪs] *n.* 骨盆

ascites [æ'saɪtɪz] *n.* 腹腔积水；腹水

ritual ['rɪtʃuəl] *n.* 仪式；惯例；礼制 *adj.* 仪式的；例行的；礼节性的

superfluous [sʊ'pɝflʊəs] *adj.* 过多的；多余的；不必要的

4. Notes to the Text A

(1) Animal welfare is the well-being of animals.

此句“well-being”是指良好的生存状态。

(2) In respect for animal welfare is often based on the belief that non-human animals are sentient and that consideration should be given to their well-being or suffering, especially when they are under the care of humans.

此句“In respect for animal welfare”指“有关动物福利方面”，“non-human animals”指“非人类的动物们”。

(3) Accordingly, some proponents argue that the perception of better animal welfare facilitates continued and increased exploitation of animals.

句中“exploitation”本意是“剥削”，此处可以理解为“人类对动物的利用”。

(4) The predominant view of modern neuroscientists, notwithstanding philosophical problems with the definition of consciousness even in humans, is that consciousness exists in nonhuman animals. However, some still maintain that consciousness is a philosophical question that may never be scientifically resolved.

此句“notwithstanding philosophical problems with the definition of consciousness even in humans”是插入语，说明在哲学方面人类关于意识的概念仍有争议，争议的焦点是只有人类才有意识，还是动物也有意识。

(5) However, some still maintain that consciousness is a philosophical question that may never be scientifically resolved.

科学是对一定条件下物质变化规律的学说，而哲学是关于世界观的学说。此句含义是指，一些观点坚持认为科学和哲学是两个范畴的问题，科学只能解决科学问题，哲学只能解决哲学问题，科学无法解决哲学问题。虽然现代神经学家认为动物是有意识存在的，但是神经学家是科学家，意识是哲学问题。

(6) This leads to further concerns about premature slaughtering such as chick culling by the laying hen industry, in which males are slaughtered immediately after hatching because they are superfluous; this policy occurs in other farm animal industries such as the production of goat and cattle milk, raising the same concerns.

在畜牧生产行业中，专业化越来越高，一种动物只用某一性别，如蛋鸡生产只有母鸡会下蛋，公鸡孵化后就淘汰了。在牛乳生产场只有母牛会产乳，生下来的小公牛立即就被淘汰了。

5. Exercises

(1) Fill in the blanks by finishing the sentences according to the passage

①Animal welfare is the ________ of animals.

②There are two forms of criticism of the concept of animal welfare, coming from ________ opposite positions.

③The predominant view of modern ________ , notwithstanding philosophical problems with the definition of ________ even in humans, is that consciousness exists in nonhuman animals.

④A major concern for the welfare of ________ animals is factory farming in which large numbers of animals are reared in ________ at high stocking densities.

⑤More ________ methods of farming, *e. g.* free range, can also raise welfare concerns such as the ________ of sheep, predation of stock by wild animals, and ________ .

⑥Farm animals are ________ selected for production ________ which sometimes impinge on the animals' welfare.

⑦While the killing of animals neednot necessarily involve ________ , the general public considers that killing an animal reduces its welfare.

(2) Answer the following questions according to the passage

①What is the standards of "good" animal welfare?

②Please give an accurate accounting of animal welfare.

③Describe the two forms of criticism of the concept of animal welfare.

④What is the predominant view of modern neuroscientists on consciousness?

⑤Explain the major concern for the welfare of farm animals.

⑥How to understand farm animals are artificially selected for production parameters which sometimes impinge on the animals' welfare?

(3) Translation of the following sentences into Chinese

①Animal welfare science uses various measures, such as longevity, disease, immunosuppression, behavior, physiology, and reproduction, although there is debate about which of these indicators provide the best information.

②Respect for animal welfare is often based on the belief that non-human animals are sentient and that consideration should be given to their well-being or suffering, especially when they are under the care of humans.

③Accordingly, some animal rights proponents argue that the perception of better animal welfare facilitates continued and increased exploitation of animals.

④Issues include the limited opportunities for natural behaviors, for example, in battery cages, veal and gestation crates, and routine invasive procedures such as beak trimming, castration, and ear notching, instead producing abnormal behaviors such as tail-biting, cannibalism, and feather pecking.

⑤The increased body weight also puts a strain on their hearts and lungs, and ascites often develops.

⑥This leads to further concerns about premature slaughtering such as chick culling by the laying hen industry, in which males are slaughtered immediately after hatching because they are superfluous; this policy occurs in other farm animal industries such as the production of goat and cattle milk, raising the same concerns.

课文A　动物福利

动物福利是动物良好生活状态。动物良好福利的标准在不同环境中有很大的差异。世界各地的动物福利团体、立法者和学者不断对这些动物福利标准进行审查、辩论、制定和修订。动物福利科学使用各种方法测量动物福利，如寿命、疾病、免疫抑制、行为、生理和生殖，这些指标中哪一项能提供最佳动物福利信息仍存争议。

关于对动物福利往往是基于一种信念，动物虽然不是人，但动物也有知觉，动物幸福或痛苦的感觉也应该考虑，特别是动物被人类管理情况下。这些关注可以包括动物是如何被宰杀作为食物的，动物如何被用于科学研究，动物如何被看护的（如宠物，动物园、农场、马戏团等地方的动物），以及人类活动如何影响野生物种的福利和生存。

关于动物福利，有两种形式的评判，二者观点完全对立。历史上一些思想家持有一种观点，认为人类对动物没有任何义务。另一种观点基于动物权利的立场，即动物不应被视为财产，而且人类对动物的任何使用都是难以接受的。因此，一些动物权利的提倡者认为，对动物福利设施更好的认识有利于增加动物的利用和持续性。因此，一些管理者将动物福利和动物权利视为两种对立的立场。另一管理者认为动物福利的增加是实现动物权利的渐进步骤。

人类关于意识的定义尽管属于哲学范畴，但现代神经学家的主流观点是动物也有意识存在。然而，仍有一些人坚持意识是一个科学永远无法解决的哲学问题。

对农场动物福利的一个主要关注点是，在工厂化生产中，大量动物被限制在高密度的圈栏中饲养的问题。其中包括动物本能的行为表现机会受到限制时，例如，用带电的笼子、犊牛和怀孕母牛笼养，以及断喙、阉割和打耳号这些有侵害性的日常操作，动物产生咬尾、相互蚕食和啄羽的异常行为。但较大范围的饲养方法，如自由放牧，也会引起福利问题，例如绵羊的割皮防蝇法、野生动物捕食家畜以及生物安全问题。

农场动物的生产是人为选择的，有时也会影响动物的福利。例如，肉鸡大规模养殖时，每只鸡都会被养得很大，以获得最大的肉产量。快速生长型肉鸡品种腿畸形的发生率高，因为大的胸肌会导致腿和骨盆发育变形，鸡的腿部不能承受其快速增加的体重，结果经常出现跛足或腿部骨折。体重的快速增加也会给其心肺带来负担，引发腹水发生。仅在英国，每年就有多达2000万的肉鸡在运达屠宰场之前，因无法承受捕捞和运输的压力而死。

另一个有关农场动物福利的问题是宰杀的方法，特别是仪式宰杀，公众一般认为杀死动物会降低动物的福利，杀死动物不必要再给动物带来痛苦。这也进一步引起了人们对成年前宰杀的关注，例如由产蛋母鸡行业淘汰的雏鸡，该行业认为小公鸡孵出后是多余的，应立即被屠宰；这种生产策略在其他动物产业也引起了同样的关注，如奶牛和奶山羊生产行业。

Part Two Extensive Reading

1. Text B Development of Animal Welfare Science

The aim of the symposium and of this issue of Acta Biotheoretica has been to bring together experts representing animal welfare science, philosophy and social sciences in order to discuss how philosophical analysis and explication of scientific terminology, concepts and theory may support the further development of animal welfare science; this is in light of the broad use of the animal welfare concept in animal ethics, society and legislation. This issue of Acta Biotheoretica is thus a contribution to the current theoretical debate on animal welfare science. The focus is on the relationship between the historical as well as practical political context and normative content of the animal welfare concept.

History shows a multi－faceted relation of animal welfare science with the established scientific disciplines as well as with animal ethics and policy. A symposium that was held in the Netherlands at the end of the 1970s on the issue of human welfare may serve as illustration. The Dutch ethologist from Groningen University Gerard Baerends was asked for a contribution from the field of animal behaviour. He started his contribution with the remark that welfare was all about subjective feelings, which he stated were however outside the realm of the study of animal behaviour in the tradition of ethology as developed by Tinbergen and colleagues (Baerends 1978). Still, he continued, ethology could contribute to the field of human welfare through its study of motivational systems as systems which strive to regain balance (within reasonable time) after challenges (homeostasis). Welfare could be framed in regaining such a balance, and "un-welfare" therefore in the impossibility to do so.

As the symposium was devoted to human welfare and Baerends was not an animal welfare researcher, the opportunity to address possible synergies between human welfare research and ethology was missed at that point. This has later become important, and Richard Haynes' and Lennart Nordenfelt's contributions to this issue specifically take a need for coherence between human and animal theories of welfare as a premise. Among the pioneers on welfare of farm animals in the early seventies, an approach similar to the biological part of Baerends' emerged: welfare was framed in terms of regaining balance, and lack of welfare in its impossibility to do so, expressed through changes in behaviour and physiology. In fact, early animal welfare research was exclusively framed in terms of chronic stress-related research, which had already established chronic stress-related symptoms indicating that animals could not reach their homeostatic condition. In the UK Donald Broom's animal welfare definition was formulated in a similar way and expanded to the various coping processes of individual organisms.

Animal welfare research for long amounted largely to measuring stress-related behaviour and physiology. This led to an uncomfortable tension with society, animal ethics and politics as animal welfare research was instigated by sympathy, if not empathy, with the poor conditions that animals were in. While politicians asked the question "Do animals suffer?", scientists replied "They show abnormal behaviour and physiology". In hindsight this focus on objective aspects of welfare might be interpreted as a direct side effect of the struggle of ethology to become accepted as a scientific discipline. Other disciplines involved in animal welfare research also keenly stuck to their standards of scientificness and to disciplinary or industrial interests when they framed animal welfare exclusive in terms of health or production parameters.

Animal welfare science only slowly gained credibility within the scientific community. From within veterinary science and animal science, animal welfare scientists seemed to criticize existing conditions and thereby carry the burden of not being independent and/or of being involved (too) emotionally. From within cognitive and neuroscience, applied researchers in contrast with laboratory researchers were criticised for having less control over experimental conditions, and from within ethology, domesticated species were seen as poor research models.

In parallel however, starting already in the early days, scientists did nevertheless take on one of the greatest challenges, that is, trying to get a grip on the question how to incorporate and measure emotions in the field of animal welfare research. By the end of the 1990s it was relatively generally accepted that sentience and emotions play an important part in welfare. Recent developments in the field of neurobiology, such as understanding cross-species brain-behaviour relationships related to reward, anxiety and stress have made it possible to address the neurobiological mechanisms of emotional processes more fruitfully. This has added to the scientific credibility of studying animals' feelings, although the relationship between the subjective experience of mental phenomena and their behavioural and neurobiological correlates or mechanisms does remain a challenge.

2. Notes to the Text B

(1) The aim of the symposium and of this issue of Acta Biotheoretica has been to bring together experts representing animal welfare science…

此句中“of the symposium and of this issue”是指进行了专题会和刊发。

(2) This is in light of the broad use of the animal welfare concept in animal ethics, society and legislation.

句中“in light of”指“根据”。

(3) The Dutch ethologist from Groningen University Gerard Baerends was asked for a contribution from the field of animal behavior.

此句中“contribution”本意为“贡献”，此处指“研究结果”。

(4) …welfare was framed in terms of regaining balance, and lack of welfare in its impossibility to do so, expressed through changes in behavior and physiology.

句中“and lack of welfare in its impossibility to do so”。这里“so”表示“regaining balance”指“重新获得平衡”，这句的意思是“如果不能重新获得平衡也就不能获得福利”。

（5） In the UK Donald Broom's animal welfare definition was formulated in a similar way and expanded to the various coping processes of individual organisms.

句中"in a similar way"是指"与早期动物福利研究同样的方法"。

（6） Other disciplines involved in animal welfare research also keenly stuck to their standards of scientificness and to disciplinary or industrial interests when they framed animal welfare exclusive in terms of health or production parameters.

此句是复合句，有"when"引导的状语从句，从句中"they"指"other disciplines involved in animal welfare research"。

（7） From within cognitive and neuroscience, applied researchers in contrast with laboratory researchers were criticised for having less control over experimental conditions, and from within ethology, domesticated species were seen as poor research models.

句中"applied researchers"是指"动物福利学研究者"。"models"是"动物模型"。

3. Answer Questions

（1） What may support the development of animal welfare science ?

（2） Why does history show a multi-faceted relation of animal welfare science with the established scientific disciplines as well as with animal ethics and policy ?

（3） Why did animal welfare science slowly gain credibility within the scientific community?

（4） How to incorporate and measure emotions in the field of animal welfare research?

课文B 动物福利的发展过程

生物理论学报的问题栏目曾召集动物福利科学、哲学和社会科学方面的专家一起，共同进行了专题讨论，即如何对科学术语、科学概念和科学理论进行哲学分析和解释，方可支持动物福利科学的进一步发展并进行了刊发；这是根据动物福利概念在动物伦理、社会和立法中的广泛运用而提出的。因此，生物理论学报对这个问题的刊发是对目前关于动物福利科学理论辩论的贡献。焦点是历史及实际政治背景与动物福利概念的规范性内容之间的关系。

历史表明，动物福利科学与既定的科学学科以及动物伦理和政策有着多方面的关系。20世纪70年代末在荷兰举行的关于人类福利问题的研讨会可以作为例证。荷兰格罗宁根大学的动物学家杰勒德·贝伦兹从动物行为方面做了研究报告。他的报告从"福利是种主观感受"开始,但他的报告有关动物行为内容超出了由汀卑尔根及同事所建立的传统动物行为学的研究范围(贝伦兹 1978)。然而,他继续说,行为学可以通过研究激励系统来为人类福利领域做出贡献,这些系统在经历挑战(体内平衡)之后(在合理的时间内),会争取再平衡。在重新获得这种平衡的过程中,福利是可以建立的,而"非福利"是不可能建立的。

正如这次研讨会对人类福利的贡献一样,贝伦兹不是一个动物福利研究人员,因此他有机会讨论人类福利研究和行为学之间可能的协同作用,这是动物福利研究的盲点。这一点后来变得很重要,理查德·海涅斯和林纳德·诺登菲尔特在这个问题上做出了贡献,特别是他们提出需要把人类福利理论和动物福利理论一致起来是研究动物福利的前提。在七十年代早期的农场动物福利中,出现了一种类似于贝伦兹的生物部分的方法:福利是以重新获得平衡为框架的,在不可能重新获得平衡时也就没有动物福利,福利表现为行为和生理的变化。事实上,早期动物福利研究完全是根据与慢性应激有关的研究进行的,这些研究已经确定了与慢性应激有关的症状,表明动物无法达到自我平衡状态。在英国,唐纳德·布鲁姆的动物福利定义也是以同样方式制定的,并扩展到了单个有机体的各种应对过程。

长期以来,动物福利研究积累了大量测定与应激相关的行为和生理指标。这些研究结果导致了社会、动物伦理和政治方面令人不安的紧张关系,因为动物福利研究是由对动物所处的恶劣环境的同情而进行的,如果不是同情动物,就不会产生社会、动物伦理和政治方面问题。当政客们问道"动物会受苦吗?"科学家回答说:"动物表现出反常的行为和生理。"事后看来,这种对福利目标方面的关注可能会被解释为行为学斗争的直接副作用,使其成为一门可接受的科学学科。其他与动物福利研究有关的学科,当他们在健康或生产参数方面限定动物福利时,也严格遵守动物福利科学标准,严格遵守动物福利学戒律或工业利益要求。

动物福利科学只是在科学界中慢慢获得了信誉。从兽医学和动物学的角度来看,动物福利科学家似乎只会批评现存的条件,无独立责任担当和/或只是感情上参与,而且参与过多。从认知和神经科学的角度来看,应用研究人员与实验室研究人员相比,实验条件的控制较少而受到批评,当从行为学角度来看,驯化物种被认为是很差的研究模型。

然而,与此同时,科学家们早期就已经开始,着手应对最大的挑战之一,即试图掌握如何在动物福利研究领域纳入和衡量情感。到20世纪90年代末，人们普遍接受知觉和情感在福利中起着重要作用。最近在神经生物学领域的发展，如有关跨物种大脑行为与奖赏、焦虑和压力间关系的研究结果，使我们能够更有成效地解决情绪过程的神经生物学机制。尽管心理现象的主观体验与其行为和神经生物学相关或神经生物学机制之间的关系仍然存挑战，但目前的这些成果，增加了研究动物情感的科学可信度。

扫码进行拓展学习

Lesson 13
Products

Part One Intensive Reading

1. Learning Objective

After learning this lesson, you should understand the following:

(1) Describe the animals which can provide the main products for humans.

(2) Explain the classification and primary role of animal by-products.

(3) Describe the differences between animal by-products and animal products.

(4) Give an overview of the main nutrients contained in meat products.

(5) Explain why meat can become main food animal products.

(6) Explain why eating livers would improve vision in dim light.

(7) Give an overview of why meat is one of the richest sources of the important B group of vitamins.

2. Text A Animal Products

An animal product is any material derived from the body of an animal. Examples are fat, flesh, blood, milk, eggs, and lesser known products, such as isinglass and rennet.

Animal by-products, as defined by the USDA, are products harvested or manufactured from livestock OTHER than muscle meat. In the EU, animal by-products (ABPs) are defined somewhat more broadly, as materials from animals that people do not consume. Thus, chicken eggs for human consumption are considered by-products in the US but not France; whereas eggs destined for animal feed are classified as animal by-products in both countries. This does not in itself reflect on the condition, safety, or "wholesomeness" of the product.

Animal by-products arecarcasses and parts of carcasses from slaughterhouses, animal shelters, zoos and veterinarians, and products of animal origin not intended for human consumption, including catering waste. These products may go through a process known as "rendering" to be made into human and non-human foodstuffs, fats, and other material that can be sold to make commercial products such as cosmetics, paint, cleaners, polishes, glue, soap and ink. The sale of animal by-products allows the meat industry to compete economically with industries selling sources of vegetable protein.

Generally, products made from fossilized or decomposed animals, such as petroleum formed from the ancient remains of marine animals, are not considered animal products. Crops grown in soil fertilized with animal remains are rarely characterized as animal products.

Several diets prohibit the inclusion of some animal products, including vegetarian, kosher, and halal. Other diets, such as veganism and the raw vegan diet, exclude any material of animal origin.

Meat is one of the main edible animal products. Perhaps most people eat meat simply because they like it. They derive rich enjoyment and satisfaction there from. For flavor, variety, and appetite appeal, meat is unsurpassed. But meat is far more than just a very temping and delicious food. Form a nutritional standpoint, it contains certain essential of an adequate diet. This is important, for how we live and how long we live determined in part by our diet.

The nutritive qualities of meats may be summarized as follow: proteins, calories, minerals and vitamins.

The word "protein" is derived from the Greek word "proteios", meaning in first place. Protein is recognized as a most important body builder. Fortunately, meat contains the proper quality and quantity of protein for the building and repair of body tissues. On a fresh basis, it contains 15% to 20% protein. Also, meat contains all of the amino acids, or building stones, which are necessary for the making of new tissue; and the proportion of amino acids in meat is similar to that in human protein.

The energy value of meat is largely dependent upon the amount of fat it contains. In turn, the fat content of meat is affected by breeding, feeding, and trimming. All of which have changed by markedly in recent years. A composite average serving of 85 g of cooked and trimmed beef, pork, lamb, or veal will provide about 10% of the recommended calorie intake for an adult.

Minerals are necessary in order to build and maintain the body skeleton and tissues and to regulate body functions. Meat is a rich source of several minerals, but it is especially good as a source of phosphorus and iron.

In 1934, Docters Minot, Murphy, and Whipple were jointly awarded the Nobel Prize in Medicine for the discovery that liver was effective in the treatment of anemia, a disease which once was regarded as fatal. The average adult would be assured an adequate supply of iron if two servings of meat were taken daily along with one serving of liver each week.

As early as 1500 B. C. the Egyptians and the Chinese hit upon the discovery that eating livers would improve one's vision in dim light. We now know that livers furnishes vitamin A, a very important factor for night vision. In fact, medical authorities recognize that night blindness, glare blindness, and poor vision in dim light are all common signs pointing to the fact that the person so affected is not getting enough vitamin A in the diet.

Meat is one of the richest sources of the important B group of vitamins, especially thiamin, riboflavin, niacin, and vitamin B_{12}. The vitamin now is used to reinforce certain foods and which are indispensable in our daily diet.

3. New Words and Phrases

flesh [fleʃ] *n.* 肉；肉体；果肉；皮肤 *v.* 用肉喂养；长胖

isinglass [ˈaɪzɪŋglæs] *n.* 鱼胶；明胶；云母

wholesomeness [ˈhəʊlsəmnɪs] *n.* 有益健康；正派；健全；生机勃勃；健康向上；有益身心

rennet [ˈrenɪt] *n.* 牛犊胃内膜；凝乳

classify [ˈklæsɪfaɪ] *vt.* 分类；归类

carcass [ˈkɑrkəs] *n.* （人或动物的）尸体；（家畜屠宰后的）躯体；骨架，遗骸，残迹等

slaughterhouses [ˈslɔːtəhaʊs] *n.* 屠宰场（等于 abattoir）；屠杀场

shelter [ˈʃɛltər] *n.* 庇护；避难所；遮盖物 *vt.* 保护；使掩蔽 *vi.* 躲避，避难

zoo [zuː] *n.* 动物园

veterinarian [ˈvɛtərəˈnɛrɪən] *n.* 兽医

cater [ˈkeɪtər] *vt.* & *vi.* 提供饮食及服务，*vt.* 满足需要，适合；投合，迎合

render [ˈrendər] *vt.* 提供；表现；使成为；宣布；翻译；回报；给予补偿；渲染，*n.* 粉刷；打底；交纳

foodstuff [ˈfudstʌf] *n.* 食品；食料

decompose [ˌdikəmˈpoz] *vt.* & *vi.* 分解；（使）腐烂

vegetarian [ˌvɛdʒəˈtɛrɪən] *n.* 素食者；食草动物 *adj.* 素食的

kosher [ˈkoʊʃər] *n.* 清洁可食的食物 *v.* 使（食物）清洁可食

halal [ˈhælæl] *adj.* 伊斯兰教律法允许的食物等；按伊斯兰教律法售卖或提供食物的

veganism [ˈvedʒənɪzəm] *n.* 纯素食主义

vegan [ˈviːgən] *n.* 严格的素食主义者

edible [ˈedəbl] *adj.* 可食用的 *n.* 食品；食物

flavor [ˈflevər] *n.* 味；韵味；特点；香料 *vt.* 给……调味；给……增添风趣

appetite [ˈæpɪtaɪt] *n.* 欲望；胃口，食欲；嗜好，爱好

unsurpassed [ˌʌnsərˈpæst] *adj.* 未被超越的；非常卓越的

delicious [dɪˈlɪʃəs] *adj.* 美味的；可口的

nutritive [ˈnjʊtrətɪv] *adj.* 有营养的；滋养的；有营养成分的；与营养有关的 *n.* 营养物

anemia [əˈnimiə] *n.* 贫血；贫血症

furnish [ˈfərnɪʃ] *vt.* 陈设，布置；提供，供应；装修（房屋）

indispensable [ˈɪndɪˈspɛnsəbl] *adj.* 不可缺少的；绝对必要的；责无旁贷的 *n.* 不可缺少之物；必不可少的人

4. Notes to the Text A

（1）Animal by-products, as defined by the USDA, are products harvested or manufactured from livestock OTHER than muscle meat.

这里的“as defined by the USDA”是插入语。USDA 是“指美国农业部”，是“United States Department of Agriculture”的首字母缩写。下句中的“EU”是指“欧盟”，是“European Union”的首字母缩写。

（2）Thus, chicken eggs for human consumption are considered by-products in the US but not France; whereas eggs destined for animal feed are classified as animal by-products in both countries.

这里“not France”省略句，完整句子为“chicken eggs for human consumption are not considered by-products in France”。“destined for”意思是指“最终用途”。

（3）These products may go through a process known as “rendering” to be made into human and non-human foodstuffs, fats, and other material that can be sold to make commercial products such as cosmetics, paint, cleaners, polishes, glue, soap and ink.

句中“known as”分词短语做定语修饰“process”。“that can be sold”定语从句修饰“material”。

(4) But meat is far more than just a very temping and delicious food.

这里“far more than”的意思是远不止于。

(5) Protein is recognized as a most important body builder.

这里“builder”是指“构成部分、构件”的意思。

(6) On a fresh basis, it contains 15% to 20% protein.

以新鲜材料为基础是化学测定过程对结果的一种表述方法，另一种表示方法是以干物质为基础。

(7) In turn, the fat content of meat is affected by breeding, feeding, and trimming. All of which have changed by markedly in recent years.

句中“of which”引导定语从句，修饰“breeding, feeding, and trimming”。“trimming”是指“修整”，目前为了提高商品感官水平，产品都有一个修整的过程。

(8) The average adult would be assured an adequate supply of iron if two servings of meat were taken daily along with one serving of liver each week.

句中“if”引导条件状语从句。“servings”本意是服务，这里是指“服用”。

5. Exercises

(1) Fill in the blanks by finishing the sentences according to the passage.

①________, as defined by the USDA, are products harvested or manufactured from livestock OTHER than muscle meat.

② The sale of animal by-products allows themeat industry to compete economically with industries selling sources of ________ protein.

③Several diets ________ the inclusion of some animal products, including vegetarian, kosher, and halal.

④________ is one of the main edible animal products.

⑤The word ________ is derived from the Greek word “proteios”, meaning in first place.

⑥The ________ value of meat is largely dependent upon the amount of fat it contains.

⑦As early as 1500 B. C. the ________ and ________ hit upon the discovery that eating livers would improve one's vision in dim light.

(2) Answer the following questions according to the passage.

①What is the difference of animal products between defined by the USDA and EU?

②Where are animal by-products from?

③What is one of the main edible animal products from the text. ?

④Why is the protein recognized as a most important body builder?

⑤Is the energy value of meat largely dependent upon the amount of fat it contains?

⑥What is now used to reinforce certain foods and which are indispensable in our daily diet?

(3) Translation of the following sentences into Chinese.

①In the EU, animal by-products (ABPs) are defined somewhat more broadly, as materials from animals that people do not consume.

②But meat is far more than just a very temping and delicious food. Form a nutritional standpoint, it contains certain essential of an adequate diet.

③Fortunately, meat contains the proper quality and quantity of protein for the building and repair of body tissues.

④In 1934, Docters Minot, Murphy, and Whipple were jointly awarded the Nobel Prize in Medicine for the discovery that liver was effective in the treatment of anemia, a disease which once was regarded as fatal.

⑤In fact, medical authorities recognize that night blindness, glare blindness, and poor vision in dim light are all common signs pointing to the fact that the person so affected is not getting enough vitamin A in the diet.

⑥Meat is one of the richest sources of the important B group of vitamins, especially thiamin, riboflavin, niacin, and vitamin B_{12}.

课文A 动物产品

动物产品是来自动物本身的任何物质。例如脂肪、肉、血液、牛乳、鸡蛋和一些不太知名的产品，如鱼胶和凝乳酶。

美国农业部有机认证定义的动物副产品是指除家畜收获或生产的肉类以外的产品。在欧盟，动物副产品（ABPs）的定义更广泛一些，主要是人们不会消费的动物材料。因此，在美国用于人类消费的鸡蛋被认为是动物副产品，而法国并不是这样认为；而用于动物饲料的鸡蛋在两个国家都被列为动物副产品。这本身并不反映产品的状况、安全性或“健康性”。

动物副产品是来自屠宰场、动物收容所、动物园和兽医的屠体部分，以及不供人食用的动物源性产品，包括餐饮垃圾。这些产品经过提取过程制成人类和非人类的食品、脂肪和其他可用于制造商业产品的材料，例如化妆品、油漆、清洁剂、抛光剂、胶水、肥皂和墨水等。动物副产品的销售使得肉类行业能够与销售植物蛋白来源的行业进行经济竞争。

一般来说，由化石或腐烂动物制成的产品，例如由古代海洋动物遗骸形成的石油，这些不被视为动物产品。在有动物遗骸的肥沃土壤中生长的作物很少被定义为动物产品。

几种饮食禁止包括一些动物产品，包括素食和清真食品。还有其他饮食，如素食主义和纯素饮食，排除任何动物来源的材料。

肉是主要的食用动物产品之一，也许大多数人吃肉只是因为喜欢。他们从中获得丰富的享受和满足感。从味道、多样性和食欲来看，肉是无与伦比的。但肉不仅仅是一种非常诱人和美味的食物，从营养学角度，它包含适当饮食的某些必要物质。这一点很重要，因为我们的生活方式以及我们的寿命决定于我们的饮食习惯。

肉类的营养价值可概括如下：蛋白质、能量、矿物质和维生素。

蛋白质这个词来源于希腊文字“proteios”，意思是第一位。蛋白质被认为是最重要的物质基础。值得一提的是，肉类含有适当的质量和数量的蛋白质用于建立和修复身体组织。如果肉质足够新鲜，它将含有15%~20%的蛋白质。此外，肉含有所有种类的氨基酸，这是制造新组织所必需的；而且肉中氨基酸的比例与人体中的蛋白质相似。

肉的能量值在很大程度上取决于它所含的脂肪量。反过来，肉类的脂肪含量也受到繁殖、饲养和修整的影响。近年来所有这些都发生了显著改变。1份约85g的熟牛肉、猪肉、羊肉或小牛肉的复合平均分量将为成人提供约10%的推荐能量摄入量。

为了构建和维持身体骨骼和组织以及调节身体功能，矿物质是必需的。肉是多种矿物质的丰富来源，尤其是作为磷和铁的来源。

1934年，Minot、Murphy和Whipple三位博士被联合授予诺贝尔生理学或医学奖，因为他们发现肝脏能有效治疗贫血，这种疾病一度被认为是致命的。如果每天吃两份肉类，并且每周一份肝脏，每个成年人基本都可以获得充足的铁质供应。

早在公元前15世纪，埃及人和中国人发现食用肝脏能够改善在昏暗的光线下的视力。我们现在知道，肝脏提供维生素 A，这是夜视非常重要的因素。事实上，医疗机构认识到患有夜盲症、眩光失明和昏暗光线下视力不佳的人们是因为在饮食中得不到足够的维生素 A。

肉是 B 族维生素最重要的来源之一，尤其是硫胺素、核黄素、烟酸和维生素 B_{12}。维生素现在被用来强化某些食物，并且在我们的日常饮食中是不可或缺的。

Part Two Extensive Reading

1. Text B How Many Things Come from Animals?

Animals play very important roles in our lives. Animals are pets, they are raised as food and they provide products important to everyday life. You may not realize how many things come from animals.

Pigs

Pork is the most widely consumed meat in the world. People eat many different pork products, such as bacon, sausage and pork chops. You might grill pork ribs in the summer, or you might enjoy a Christmas ham. A 113 kg market hog yields about 68 kg of pork. In addition to pork, several valuable products come from pigs. These include insulin for the regulation of diabetes, valves for human heart surgery, suede for shoes and clothing, and gelatin for foods and non-food uses. Pigs by-products are also important parts of products such as water filters, insulation, rubber, antifreeze, certain plastics, floor waxes, crayons, chalk, adhesives and fertilizer. Lard is fat from pig abdomens and is used in shaving creams, soaps, make-up, baked goods and other foods.

Cattle

"Beef: It's what for dinner" is a phrase that you might hear in commercials. So what products do beef animals give us? Beef cattle provide different cuts of meat that many of us eat every day. These include ribs, steak and ground beef. Dairy cows are another type of cattle that provide us with nutritional products. There are many different dairy products but some you might be most familiar with include milk, cheese, yogurt, butter and ice cream. Other dairy products include sour cream, cottage cheese, whey, cream cheese and condensed milk. Dairy products are often used in cooking and baking and contain calcium, which can help to strengthen your bones.

When dairy animals can no longer produce milk, they are often used for meat, primarily in the form of ground beef. It is possible to get the same cuts of meat from a dairy cow that you do a beef animal. Male dairy calves, called bull calves, that are not used for beef are often used for veal. Veal is meat from younger animals and is very lean. It is considered a delicacy in many countries.

There are various by-products that come from all types of cattle. Tallow is fat from cattle, and it is used in wax paper, crayons, margarine, paints, rubber, lubricants, candles, soaps, lipsticks, shaving creams and other cosmetics.

Gelatin is a protein obtained by boiling skin, tendons, ligaments, and/or bones of cattle in water. Gelatin is used in shampoos, face masks and other cosmetics. Gelatin is also used in foods as a thickener for fruit gelatins and puddings, candies and marshmallows.

Leather comes from the hides of animals. It is used to make wallets, purses, furniture, shoes and car upholstery. Leather can be made from the skin of pigs, cattle, sheep, goats and exotic species such as alligators.

2. Notes to the Text B

When dairy animals can no longer produce milk, they are often used for meat, primarily in the form of ground beef.

句中“when”引导的状语从句，意思是当不再产乳的时候。

3. Answer Questions

(1) Can you specify s come from animals?

(2) Is pork the most widely consumed meat in the world?

(3) What's products do beef animals give us?

(4) Which meat form is come from chickens?

课文B　你知道有多少东西来自动物吗?

动物在我们的生活中扮演非常重要的角色。动物是宠物，它们是作为食物饲养的，它们提供对日常生活重要的产品。你可能还没有意识到有多少东西来自动物。

猪

猪肉是世界上消费量最大的肉类。人们吃许多不同的猪肉产品，例如培根、香肠和猪排。你可以在夏天烤猪排，也可以享用圣诞火腿。市场上113kg的猪可以产约68kg的猪肉。除猪肉之外，还有几种有价值的产品来自猪。这些包括用于调节糖尿病的胰岛素，用于人类心脏手术的瓣膜，用于鞋子和衣服的绒面革以及用于食品和非食品用途的明胶。猪副产品也是产品的重要组成部分，例如水过滤器、隔热材料、橡胶、防冻剂、某些塑料、地板蜡、蜡笔、白垩、黏合剂和肥料，猪油是来自猪腹部的脂肪，用于剃须膏、肥皂、化妆品、烘焙食品和其他食品。

牛

“牛肉：你的晚饭”可能是你在商业广告中听到的一句话。那么牛这种动物带给我们的产品是什么？肉牛为我们提供了许多人每天都要吃的不同肉类，包括排骨、牛排和碎牛肉。奶牛是另一种为我们提供营养品的牛。有许多不同的乳制品，但你可能最熟悉的包括牛乳、干酪、酸乳、黄油和冰淇淋，其他乳制品包括酸奶油、干奶酪、乳清、奶油芝士和炼乳。乳制品经常用于烹饪和烘焙，且其中含有钙，这可以帮助强化骨骼。

当奶牛不能再生产牛乳时，它们通常用于肉类，主要以碎牛肉的形式。从奶牛那里也可以得到同样的肉。雄性奶牛犊常常用于小牛肉。小牛肉来自年幼动物，非常瘦，它在许多国家被认为是美味佳肴。

有很多来自各种牛的副产品。牛脂是牛的脂肪，用于蜡纸、蜡笔、人造黄油、油漆、橡胶、润滑剂、蜡烛、肥皂、口红、剃须膏和其他化妆品。

明胶是一种蛋白质，通过将牛的皮肤、肌腱、韧带或骨头在水中煮沸而获得。明胶用于香波、面膜和其他化妆品。明胶也用于食品中作为水果明胶和布丁、糖果和棉花糖的增稠剂。

皮革来自动物的皮。它用于制作钱包、手提包、家具、鞋子和汽车内饰。皮革可以由猪、牛、绵羊、山羊和鳄鱼等外来物种的皮制成。

扫码进行拓展学习

Unit Ⅳ Animals Care

Lesson 14

Beef Cattle

Part One Intensive Reading

1. Learning Objective

After learning this lesson, you should understand the following:

(1) Describe how to provide necessary food, water and care to protect the health and well-being of animals.

(2) Give an overview of how to protect herd health, including access to veterinary medical care.

(3) Explain what facilities should be provided that allow safe, humane, and efficient movement and/or restraint of cattle.

(4) Describe how to use appropriate methods to humanely euthanize terminally sick or injured livestock and dispose of them properly.

(5) Describe how to provide personnel with training/experience to properly handle and care for cattle.

(6) Give an overview of how to make timely observations of cattle to ensure basic needs are being met.

(7) Describe how to minimize stress when transporting cattle.

(8) Explain how to keep updated on advancements and changes in the industry

to make decisions based upon sound production practices and consideration for animals well-being.

(9) Give an overview of what's persons willfully mistreat animals will not be tolerated.

2. Text A Feeding Guidelines for Beef Cattle

Cattlemen have long recognized the need to properly care for livestock. Sound animals husbandry practices, based on decades of practical experience and research, and are known to impact the well-being of cattle, individual animals health and herd productivity. Cattle are produced in very diverse environments and geographic locations in the United States. There is not one specific set of production practices that can be recommended for all cattle producers. Personal experience, Beef Quality Assurance (BQA) training and professional judgment can serve as a valuable resource for providing proper animal care. The following information is to be used as an educational resource, all production practices should be adapted to specific needs of individual operations.

Feeding and nutrition

Diets for all classes of beef cattle should meet the recommendations of the National Research Council (NRC) and/or recommendations of a nutritional consultant. For local recommendations and advices, contact your state agricultural extension as a potential resource.

- Cattle must have access to an adequate water supply. Estimated water requirements for all classes of beef cattle in various production settings are described in the National Academy of Sciences NRC Nutrient Requirements of Beef Cattle.
- Provide adequate feed. Avoid feed and water interruption longer than 24 hours.
- Feedstuffs and feed ingredients should be of satisfactory quality to meet nutritional needs.
- Under certain circumstances (*e.g.*, droughts, frosts, and floods), test feedstuffs or other dietary components to determine the presence of substances that can be detrimental to cattle well-being, such as nitrates, prussic acids, mycotoxins, etc.
- Producers should become familiar with potential micronutrient deficiencies or excesses in their respective geographical areas and use appropriately formulated supplements.

• Use only USDA, FDA and EPA approved products for use in cattle. These products must be used in accordance with the approved product use guidelines.

Feeding guidelines for beef cows

Body condition scoring of beef cows is a scientifically approved method to assess nutritional status. Body condition scores (BCS) range from 1 (emaciated) to 9 (obese).

• A BCS of 4 ~ 6 is most desirable for health and production. A BCS of 2 or under is not acceptable and immediate corrective action should be taken.

• During periods of prolonged drought and widespread shortages of hay and other feedstuffs, the average BCS of cows within a herd may temporarily decline. This is not desirable, but may be outside the cattle owner's control until drought relief is achieved.

• During periods of decreasing temperature, feeding plans should reflect increased energy needs.

Feeding Guidelines for Stocker Cattle

• Stockers are raised on a wide variety of forages (native pasture, annuals, improved pasture) with minimal additional nutrient supplementation.

• On growing forages, stocking rates should be established that meet production goals for growth and performance.

• On dormant pastures, supplement cattle as needed to meet maintenance or growth requirements for the animal's weight, breed, and age as established by NRC guidelines and targeted production goals of the operation.

Feeding guidelines for feeder cattle

Feedyard cattle can eat diverse diets, but the typical ration contains a high proportion of grain (s) (corn, milo, barley, grain by-products) and a smaller proportion of roughages (hay, straw, silage, hulls, etc.). The NRC lists the dietary requirements of beef cattle (based on weight, weather, frame score, etc.) and the feeding value of various commodities included in the diet.

• Consult a nutritionist (private consultant, university or feed company employee) for advice on ration formulation and feeding programs.

• Avoid sudden changes in ration composition or amount of ration offered.

• Monitor changes in weight gain, feces, incidence of digestive upsets (acidosis or bloat) and foot health to help evaluate the feeding program.

• A small percentage of cattle in feedyards develop laminitis or founder.

Mild cases do not affect animal welfare or performance; however, hooves that are double their normal length compromise movement. In these instances, the individual animal should be provided appropriate care and marketed as soon as possible.

3. New Words and Phrases

husbandry [hʌzbəndri] *n.* 畜牧业，饲养业

impact [ɪmˌpækt] *n.* 影响；碰撞；*vt.* 撞击；挤入，压紧；对……产生影响；*vi.* 产生影响；冲撞

productivity [prɑːdʌkˈtɪvəti] *n.* 生产率，生产力

geographic [dʒiːəˈgræfɪk] *adj.* 地理学的，地理的

recommendation [rɛkəmɛnˈdeʃən] *n.* 推荐；建议；推荐信；可取之处

consultant [kənˈsʌltənt] *n.* （受人咨询的）顾问；会诊医生

adequate [ˈædɪkwət] *adj.* 足够的；适当的，恰当的；差强人意的；胜任的

interruption [ɪntəˈrʌpʃən] *n.* 中断；打断；障碍物；打岔的事

ingredient [ɪŋˈgriːdɪənt] *n.* （烹调的）原料；（构成）要素；因素

circumstance [ˈsɜːrkəmstæns] *n.* 环境，境遇；事实，细节；典礼，仪式

frost [frɔst] *n.* 霜冻，严寒天气

nitrate [ˈnaɪˌtretˌ-trɪt] *n.* 硝酸盐；硝酸根；硝酸酯；硝酸盐类化肥

prussic acid [ˈprʌsɪk ˈæsɪd] *n.* 氰酸，氢氰酸

mycotoxin [ˌmaɪkoʊˈtɒksən] *n.* 霉菌毒素

prolonged [prəˈlɔːŋd] *adj.* 延长的；持续很久的；拖延的

temporarily [tempəˈrerɪlɪ] *adv.* 暂时地；临时地

procedures [prəˈsiːdʒəz] *n.* 程序；手续；手续；步骤；常规

stocker cattle 食用牛，候宰牛

forage [ˈfɔːrɪdʒ] *n.* 牛马饲料

dormant [ˈdɔːrmənt] *adj.* 潜伏的，蛰服的，休眠的；静止的

roughages [ˈrʌfɪdʒ] *n.* 粗饲料，粗粮，粗糙的原料

strawn [strɔ] *n.* 稻草；吸管；麦秆；毫无价值的东西 *adj.* 稻草的，麦秆的；稻草做的；假的，假想的；无价值的

laminitis [ˌlæməˈnaɪtɪs] *n.* 蹄叶炎；板炎

hooves [huːvz] *n.* hoof 的复数；（兽的）蹄，马蹄

4. Notes to The Text A

(1) The following information is to be used as an educational resource, all production practices should be adapted to specific needs of individual operations.

此句主要学习短语“be adapted to”，在文中的意思是“适合于、适应”。

（2）Estimated water requirements for all classes of beef cattle in various production settings are described in the National Academy of Sciences (NRC) Nutrient Requirements of Beef Cattle.

此句主要认识短语“all classes of”指“各个等级”。“are described in”指“被描述的”。

（3）On growing forages, stocking rates should be established that meet production goals for growth and performance.

句中“stocking rates”在这里的意思是“载畜率、放养率”。

5. Exercises

（1）Fill in the blanks by finishing the sentences according to the passage.

①Cattlemen have long ________ the need to properly care for livestock.

②There is not one specific set of production practices that can be ________ for all cattle producers.

③Feedstuffs and feed ingredients should be of ________ quality to ________ meet nutritional needs.

④These products must be used in ________ with the approved product use guidelines.

⑤Avoid sudden changes in ration ________ or amount of ration offered.

（2）Answer the following questions according to the passage.

①What can be a valuable resource for providing proper animals care?

②What can be determined the presence of substances that can be detrimental to cattle well-being under certain circumstances?

③What is the most desireable BCS for beef health and production?

④What are the NRC lists the dietary requirements of beef cattle and the feeding value of various commodities included in the diet?

⑤ How does the hooves that are double their normal length compromise movement?

（3）Translation of the following sentences into Chinese.

①Sound animals husbandry practices, based on decades of practical experience and research, and are known to impact the well-being of cattle, individual animals health and herd productivity.

② Personal experience, Beef Quality Assurance (BQA) training and professional judgment can serve as a valuable resource for providing proper animal care.

③Producers should become familiar with potential micronutrient deficiencies or excesses in their respective geographical areas and use appropriately formulated supplements.

④Stockers are raised on a wide variety of forages (native pasture, annuals, improved pastures) with minimal additional nutrient supplementation.

⑤Diets for all classes of beef cattle should meet the recommendations of the National Research Council (NRC) and/or recommendations of a nutritional consultant.

课文 A　肉牛饲养指南

养牛户早就意识到，必须妥善照料牲畜。根据多年的实践经验和研究成果，已知采取合理的畜牧业措施会促进动物的福利、健康和群体生产力。牛在美国多种多样的环境和地理区域均有饲养，所以没有一套固定的生产方法可以推荐给所有的养牛者。个人经验、牛肉质量保证（BQA）培训和专业判断均可提供适合动物管理的宝贵资料。以下的信息将用作教育资源，所有生产实践应能适合个体业务的具体需要。

饲养与营养

肉牛的各种日粮都应符合美国国家研究委员会（NRC）的建议和/或营养顾问的建议。为了得到当地的建议和意见，联系你所在州的农业推广机构，作为一个潜在的资源。

- 必须为牛提供足够的饮水。在美国国家研究委员会 NRC 肉牛营养需要量中，给出各种生产环境中各类肉牛的估计需水量。
- 提供足够的饲料。避免供料及供水中断超过 24h。
- 饲料的品质应令人满意，饲料成分满足营养需要。
- 在一些情况下（例如干旱、霜冻和洪水），检测饲料或其他日粮成分，以确定是否存在有害牛健康的物质，例如硝酸盐、氢氰酸、霉菌毒素等。
- 生产者应熟悉其地理区域可能存在的微量营养素缺乏或过量，并选用适当的补充剂。
- 只选用美国农业部（USDA）、美国食品药品监督管理局（FDA）和美国国家环境保护局（EPA）批准的用于牛的产品。这些产品必须按照推荐的产品使用指南使用。

母牛饲养指南

母牛体况评分法是一种经科学认可的评定营养状况的方法。体况评分（BCS）范围从1（消瘦）到9（肥胖）。

● 母牛健康和生产最理想的体况评分是4~6。体况评分2或2以下是不能接受的，应立即采取纠正措施。

● 在长期干旱和干草及其他饲料普遍短缺的时期，牛群中牛的平均体况评分可能会暂时下降。这是人们不希望的，但在干旱缓解之前，牧场主人可能无法控制。

● 在气温下降期间，饲养应反映能量需求的增加。

架子牛饲养指南

● 架子牛饲喂多种牧草（天然牧草、一年生草本、改良牧草），尽量减少额外营养补充剂。

● 对于牧草生长季节，应确定合适的放养率，确保生长和体型满足生产目标。

● 对休眠期牧场，NRC指南和企业生产目标根据不同品种、年龄、体重给出了维持和生长需要的营养补充量。

育肥牛饲养指南

牛场的牛可以食用不同的饲料，但典型的日粮含有较高比例的谷物（玉米、高粱、大麦、谷物副产品）和较小比例的粗饲料（干草、稻草、青贮饲料、谷壳等）。NRC列出了肉牛的日粮需求（基于体重、天气、体型评分等），同时给出各种日粮的饲养价值。

● 咨询营养学家（私人顾问、大学或饲料公司雇员），获得有关日粮配方和喂养计划的建议。

● 避免突然改变日粮组成或日粮喂量。

● 监测增重、粪便、消化不良（酸中毒或腹胀）发生率及蹄部健康状况的变化，以帮助评估喂养计划。

● 饲养场的一小部分牛会发生蹄炎或发育不良。

轻微的病例不会影响动物的福利或生产性能，但是，蹄长超过正常长度一倍会影响运动。在这些情况下，应进行单独饲养并尽快出售。

Part Two Extensive Reading

1. Text B Beef Cattle

Shelter and housing

Cattle in backgrounding facilities or feedyards must be offered adequate space for comfort, socialization and environmental management.

- Pen maintenance, including manure harvesting, will help improve pen conditions.
- Mud is more of a problem in the winter with low evaporation rates or improper drainage conditions. Accumulation of mud on cattle should be monitored as a measure of pen condition and cattle care in relation to recent weather conditions.
- Feedyards should use dust reduction measures to improve animals performance.
- Floors in housing facilities should be properly drained and barns and handling alleys should provide adequate traction to prevent injuries to animals and handlers.
- Handling alleys and housing pens should be free of sharp edges and protrusions to prevent injury to animals and handlers.
- Design and operate alleys and gates to avoid impeding cattle movement. When operating gates and catches, reduce excessive noise, which may cause distress to the animals.
- Adjust hydraulic or manual restraining chutes to the appropriate size of cattle to be handled. Regular cleaning and maintenance of the working parts is imperative to ensure the system functions properly and is safe for the cattle and handlers.
- Mechanical and electrical devices used in housing facilities should be safe.

2. Notes to the Text B

(1) Handling alleys and housing pens should be free of sharp edges and protrusions to prevent injury to animals and handlers.

认识短语：“be free of” 意为“免于，没有”。

(2) If more than 25% of cattle jump or run out of the chute, there should be a review of the situation and questions asked such as: is this a result from cattle temperament or prior handling issue, was the chute operating properly, etc.?

句中“If”引导条件状语从句。

(3) Regular cleaning and maintenance of the working parts is imperative to ensure the system functions properly and is safe for the cattle and handlers.

认识短语：“be imperative to”意为“有必要的”。

3. Answer Questions

(1) What must cattle in backgrounding facilities or feedyards be offered adequate space for?

(2) What is more of a problem in winter for beef?

(3) What is the measure offeedyards to improve animals performance?

课文B 肉牛

牛棚和牛舍

牛棚和牛舍必须为设施圈养或饲养场的牛群提供足够的舒适、社会化和环境管理空间。

- 围栏饲养，包括粪肥清除，将有助于改善围栏条件。
- 在蒸发率低或排水条件不当的冬季，泥浆是个大问题。应根据最近的天气情况，结合监测牛身上的泥土沾污，评估圈养条件和牛的饲养情况。
- 饲养场可采取减少尘埃的措施，以改善动物生产性能。
- 房屋设施的地板应妥善排水，牛舍和过道坚实，以防止动物和操作人员受伤。
- 过道和围栏应避免尖锐和边缘突出，以免伤害动物和饲养员。
- 设计操作过道和大门，以避免妨碍牛的活动。在操作闸门和捕捉器时，减少过多的噪声，因为这些噪声可能会伤害动物。
- 调校水压或水槽，使其适合牛只。定期清洁和维修工作部件，确保系统正常运作，确保牛只和饲养员是安全的。
- 房屋中使用的机械和电气设备在任何情况下都应确保安全。

扫码进行拓展学习

Lesson 15
Dairy Cattle

Part One Intensive Reading

1. Learning Objective

After learning this lesson, you should understand the following:

(1) Describe what special provisions to transport with.

(2) Explain what is non-ambulatory animals and lameness.

(3) Give an overview of what might be responsible for causing feet and leg problems.

(4) Describe producer actions: prevent, detect, treat, cull and euthanize.

(5) Explain why this guide is designed to.

(6) Explain why early recognition of problems and prompt, appropriate treatment are key factors in preventing the loss of an animal.

(7) Describe what producers are encouraged to work with their herd veterinarians for.

2. Text A Transport of Special Provisions

Animals that are to be transported with special provisions are not to be transported to sales or auction barns or collection yards. Selling these animals to a dealer who will then transport them to a sales barn, auction barn or collection yard is unacceptable. Special provisions could include extra bedding or segregating them on the truck to ensure their welfare and comfort during transit.

In addition to the cases outlined, producers must assess each animal based on its individual state of health prior to making a decision to load or not, and whether it should go to an auction or directly to a processing plant. Animals must only be loaded if they are assessed to be fit at the farm and able to withstand the journey to its destination.

Non-Ambulatory Animals and Lameness

Non-ambulatory animals (sometimes referred to as "downers") are those unable to get up, walk or remain standing without assistance. Animals may become downers from an obvious physical problem, such as a broken leg, or from weakness caused by emaciation, dehydration, exhaustion or disease.

Leg problems in cattle can be caused by a variety of factors including fractures, abscesses, arthritis, laminitis and foot rot. The entire animal should be assessed, as a lame animal in poor body condition will likely be condemned at the processing facility.

Feet and leg problems can result in poor performance and substantial economic loss. Several factors might be responsible for causing problems:

- Nutrition and feeding practices
- Facility and physical environment
- Genetic predisposition
- Other ongoing diseases

A lame animal can only be transported if it can rise, stand and walk under its own power. Use the following lameness classes to determine the best option when dealing with sick or injured cows and calves.

Lameness classes

Class 1: Visibly lame but can keep up with the group; no evidence of pain.

Class 2: Unable to keep up; some difficulty climbing ramps. Load in rear compartment.

Animals in lameness classes 1 and 2 can be transported directly to slaughter or to a veterinary clinic for treatment. Segregate and load class 1 and 2 animals in rear compartments with ample bedding.

Class 3: Requires assistance to rise, but can walk freely.

Class 4: Requires assistance to rise; reluctant to walk; halted movement.

Class 5: Unable to rise or remain standing.

Do not load or transport class 3, 4 or 5 animals except for veterinary treatment and with the advice of a veterinarian.

Producer actions include prevent, detect, treat, cull, euthanize

①Keep accurate records of all animals.

②Improve poor facility design, such as lying, walking and loading surfaces.

③Hoof trim and/or evaluate feet at least once per year.

④Cull animals with persistent problems.

⑤Assess the risk of an animal becoming non-ambulatory in transport before loading the animal.

⑥Provide prompt medical care in consultation with your veterinarian.

⑦Euthanize animals in lameness class 3, 4 and 5.

⑧Emergency on farm slaughter if animal is fit for human consumption and under 30 months of age.

Abscess

An abscess is a localized collection of pus in a cavity of disintegrated tissue. Some minor abscesses can be treated on farm. Multiple abscesses may be caused by a major illness involving other portions of the body and may result in condemnation of the carcass at slaughter.

- It should be noted that transport animals with minor abscesses directly to slaughter, and do not load or transport animals with multiple abscesses.

Producer actions are as follows.

①Check animals for abscesses regularly and treat affected animals as soon as possible.

②Try to identify source if multiple abscesses are present, in consultation with veterinarian.

③Provide prompt medical care in consultation with your veterinarian.

④Euthanize animals with multiple abscesses.

Arthritis

Arthritis is an inflammation of the joint, characterized by a progressive difficulty moving and increased time spent lying down with the affected joints flexed. Swollen joints can be a symptom of arthritis. Treatment is dependent on the degree of lameness. Two or more affected joints can cause an animal to be condemned at slaughter.

Animals should be assessed according to the lameness class 1 through 5. Animals with arthritis inmultiple joints or animals that are judged to be in lameness classes 3, 4 or 5 should not be transported.

Producer Actions are as follows.

①Observe all cows and calves for swollen joints.

②Determine cause if several animals are affected.

③Detect and treat early or ship promptly.

④Provide prompt medical care in consultation with your veterinarian.

⑤Euthanize animals in lameness classes 3, 4 and 5.

⑥Emergency on farm slaughter if the animal is fit for human consumption and under the age of 30 months.

3. New Words and Phrases

non-ambulatory [ˈnəʊn] [ˈæmbjələtɔːri] *adj.* 非走动的，非流动的

intervention [ˌɪntəˈvenʃn] *n.* 介入，干涉，干预；调解，排解

provision [prəˈvɪʒən] *n.* 规定，条项，条款；预备，准备，设备；供应，（一批）供应品；生活物质，储备物资

lameness [leɪmnəs] *n.* 跛行；跛，残废，僵而疼痛

emaciation [ɪˌmeɪsiːeɪʃn] *n.* 消瘦，憔悴，衰弱

dehydration [ˌdihaɪˈdreʃən] *n.* 脱水；失水；干燥，极度口渴

exhaustion [ɪgˈzɔstʃən] *n.* 疲惫，衰竭；枯竭，用尽；排空；彻底的研究

abscess [ˈæbses] *n.* 脓肿 *vi.* 形成脓肿 *adj.* 形成脓肿的

arthritis [ɑːrˈθraɪtɪs] *n.* 关节炎

predisposition [ˌpridɪspəˈzɪʃən] *n.* 倾向，素质；易染病体质

euthanize [ˈjuːθənaɪz] *vt.* 使安乐死，对……施无痛致死术

condemnation [ˌkɑːndemˈneɪʃn] *n.* 谴责；定罪；谴责（或定罪）的理由；征用

symptom [ˈsɪmptəm] *n.* 症状；征兆

4. Notes to the Text A

（1）Unfortunately, the reality is that some animals will become injured or sick to the extent that they are considered unfit, compromised or at risk.

此句中有两个“that”，第一个“that”引导的是表语从句，第二个“that”从句引导同位语从句。

（2）This guide is designed to assist dairy and beef producers to recognize health-related problems and respond to them in a timely and responsible manner.

认识短语：“respond to”意为“响应，对……做出反应”；“in a…manner”意为“以……方式”。

（3）Animals that are to be transported with special provisions are not to be transported to sales or auction barns or collection yards.

认识短语：“be transported with”意为“被运输”。“that”引导定语从句。

（4）In addition to the cases outlined producers must assess each animal based on its individual state of health prior to making a decision to load or not, and whether it should go to an auction or directly to a processing plant.

认识短语：“in addition to” 意为“除……之外”。“Whether” 引导同位语从句。

（5） Multiple abscesses may be caused by a major illness involving other portions of the body and may result in condemnation of the carcass at slaughter.

句中 “involving” 现在分词，表示经常主动的动作。

5. Exercises

（1） Fill in the blanks by finishing the sentences according to the passage.

①Feet and leg problems can result in poor ________ and substantial economic loss.

②Feet and leg problems can result in poor ________ and substantial economic loss.

③The entire animal should be assessed, as a lame animal in poor body condition will likely be ________ at the processing facility.

④Assess the risk of an animal becoming ________ in transport before loading the animal.

⑤Two or more affected joints can cause an animal to be ________ at slaughter.

（2） Answer the following questions according to the passage.

①What are the special provisions should include during animal transit?

②What are the non-ambulatory animals referred to?

③How to define the animals become downers?

④What are the factors might be responsible for causing animal problems?

⑤What's the dairy cattle arthritis characterized by?

（3） Translation of the following sentences into Chinese.

①This would include animals that are non-ambulatory (downers), unable to stand without assistance or to move without being dragged or carried.

②Producers are encouraged to work with their herd veterinarians for early intervention treatment and culling decisions.

③Animals must only be loaded if they are assessed to be fit at the farm and able to withstand the journey to its destination.

④Emergency on farm slaughter if animal is fit for human consumption and under 30 months of age.

⑤Animals with arthritis in multiple joints or animals that are judged to be in lameness classes 3, 4 or 5 should not be transported.

课文 A　特殊规定的运输

按照特殊规定运输的动物，不得运往市场、拍卖场或者收集场。也不能把这些动物卖给经销商，再由经销商把它们运到销售谷仓、拍卖谷仓或收集场。特别规定还包括运输车辆的床位或者隔离设施，要确保它们在运输过程中的福利和舒适。

除上述情况外，生产者在决定是否装车之前，还必须根据每只动物的健康状况对其进行评估，并决定是否将其送往拍卖或直接送往加工厂。动物必须经过农场的评估，确定它们适合在农场里生活，并且能够经受住旅程抵达目的地才能被装运。

无行动能力的动物和跛行的动物

无行动能力的动物（有时被称为“不能站立”）是指那些在没有协助的情况下无法起身、行走或站立的动物。动物可能因为明显的身体问题，如腿部骨折，或者因为消瘦、脱水、疲惫或疾病引起的虚弱而变成无行动能力的动物。

牛的腿部问题可能是由多种因素引起的，包括骨折、脓肿、关节炎、蹄叶炎和足部腐烂。应该对整只动物进行评估，因为身体状况不佳的跛脚动物很可能会被送到加工厂处理。

蹄和腿的问题可能会导致生产性能不佳和严重的经济损失。多种因素可能会引起此问题：

- 营养和喂养方法
- 设施和物理环境
- 遗传易感性
- 其他正在发生的疾病

只有跛足的动物能自行站起来，自行走路，才能被运送。在处理生病或受伤的母牛和小牛时，使用下面的程序评价跛行程度，确定最佳处理方法。

跛行分级

1 级：明显跛足，但能跟上牛群；没有疼痛的表现。

2 级：跟不上牛群；爬坡有困难。后躯负重。

第 1、2 级跛行动物可以直接送到屠宰场或兽医诊所进行处理。第 1、2 级跛行动物也可以装入有充足床位的后隔间进行隔离。

3 级：跛行动物需要扶助才能起身，但可以自由行走。

4 级：跛行动物需要扶助才能起身；不愿意行走；停止活动。

5 级：跛行动物不能起身，不能站立。

不能装载或运送的第3、4或5级动物，除非接受兽医治疗和兽医的建议。

生产者可以进行的操作如下所示，包括预防、诊断、治疗、淘汰、安乐死。

①保存所有动物的准确记录。

②改善不良设施的设计，如躺卧、行走和负重表面。

③修蹄和/或至少每年评估脚一次。

④淘汰有顽固问题的动物。

⑤在装载动物之前，评估动物在运输过程中变得不能行走的风险。

⑥咨询您的兽医提供及时的医疗服务。

⑦对跛行3、4、5级的动物实施安乐死。

⑧适宜供人食用而年龄未满30个月的动物，进行紧急农场屠宰。

脓肿

脓肿是一种局限性的脓液聚集在已经分解的组织腔内。一些小脓肿可以在农场治疗。多发性脓肿可能是由身体其他部位的重大疾病引起的，并可能影响屠宰时对胴体的评价。

- 把有轻微脓肿的动物直接送去屠宰。
- 不要装载或运送多发性脓肿的动物。

生产者可以进行的操作如下所示。

①定期检查动物是否有脓肿，并尽快对受感染的动物进行治疗。

②咨询兽医找出多发性脓肿的来源。

③咨询兽医，确定及时的医疗服务。

④对多发性脓肿的动物实施安乐死。

关节炎

关节炎是关节的一种炎症，症状是运动逐渐变得困难，需要弯曲关节躺下的时间也会增加。关节肿胀可能是关节炎的症状。处置方式取决于跛行的程度。两个或两个以上受影响的关节可能会导致动物被定为进入屠宰场。

动物应根据跛行程度1~5级进行评估。多关节有关节炎的动物，或被判定为跛行等级3、4或5的动物，不应运输。

生产者可以进行的操作如下所示。

①观察所有母牛和小牛的关节肿胀。

②确定动物受到影响的原因。

③及早发现并及时治疗。

④与兽医协商提供及时的医疗服务。
⑤对3、4、5等级的跛行动物实施安乐死。
⑥适宜供人食用及年龄不足30个月的动物，须进行紧急农场屠宰。

Part Two Extensive Reading

1. Text B Disease Control of Dairy Cattle

Displaced Abomasum

A displaced abomasum (twisted stomach) is a repositioning of the fourth stomach from its normal position on the bottom of the abdomen to the upper left side in most cases. It occurs most frequently in high-producing, heavily fed dairy cattle. One of the chief symptoms is a sudden or gradual decrease in appetite. Other symptoms include scanty bowl movements, soft and discoloured with some occasional diarrhea.

- Transport animals directly to slaughter with special provisions in separate compartment with adequate bedding.
- Transport only if animals are not showing signs of weakness, dehydration or pain (*e. g.* grinding teeth, arched back).

Producer actions are as follows.

①Preventable by dietary adjustment.
②Provide prompt medical care in consultation with your veterinarian.
③Consult veterinarian to distinguish from ketosis.
④Treatable by surgery.

2. Notes to the Text B

A displaced abomasum (twisted stomach) is a repositioning of the fourth stomach from its normal position on the bottom of the abdomen to the upper left side in most cases.

剖析句子：“from…to”指“从……到……”，是对“A displaced abomasum”的解释。

3. Answer Questions

(1) How does the displaced abomasum most frequently occur in dairy cattle?

(2) What's the chief symptoms of displaced abomasum?

(3) How to reduce the incidence of displaced abomasum in dairy herd?

课文B 奶牛疾病控制

皱胃移位

皱胃移位（扭曲的胃）是指大多数情况下第四胃从腹部底部的正常位置移向左上方的位置。这种病最常发生在高产、过度喂料的奶牛。主要症状之一是食欲突然或逐渐下降。其他症状包括少量的回头望腹、偶尔腹泻、拉白色软便。

需要注意的是可以把动物直接运往屠宰场的隔离间，里面备有特别设施且卧床足够。只有在动物没有出现虚弱、脱水或疼痛的症状（如磨牙、弓背）时，才可运送动物。

生产者可以进行的操作如下所示。

①通过饮食调整可以预防。

②与兽医协商提供及时的医疗。

③咨询兽医以区别酮症。

④可以通过手术治愈。

扫码进行拓展学习

Lesson 16

Pig

Part One Intensive Reading

1. Learning Objective

After learning this lesson, you should understand the following:

(1) Define the concept of parity segregation and give an overview of the reason of parity segregation caused.

(2) Describe what is the extra cost factors we must consider for parity segregation.

(3) Give the big difference in micro-mineral intakes between the first two pregnancies and those of the more mature sows.

(4) Explain why the best advice for the time being is to understand the mechanisms of the concept and do all you can to minimise contact between the sows in the first two parities from those in subsequent ones and likewise for their offspring in the nurseries.

(5) Give an overview of the future of parity segregation.

2. Text A Parity Segregation

This development should be of particular interest to pig breeders due to growing evidence of its influence on disease. While it has been discussed by academics for at least 5 years, only within the past 2 years has attention been drawn to the concept, almost exclusively in the USA by articles and at technical meetings. Pig breeders are vague about what it entails and have assumed that it means raising sows separately, parity by parity.

Because the gilt and possibly the young P1 sow, and even the P2 sow, are such a source of infection due to their underdeveloped immune systems, medication costs are reduced in the more mature sows and their offspring by up to 50%, and a

20% reduction looks to be typical. Additionally there is a evidence that Parity Segregation is much more likely to extend the sow productive life of the herd (SPL).

A future area of research associated with the concept could be that the progeny of young P1 and P2, even possibly P3 (sows if these latter are from a highly productive strain of female) could benefit from a different balance and/or amount of nutrients compared to the progeny of the more mature P4 to P7～P10 sows. Big difference in micro-mineral intakes between the first two pregnancies and those of the more mature sows. If this is typical surely we can no longer afford to feed the same gestation diets to at least the P0 and P1 sows (and probably the P2 sows as well) compared to the rest of the herd which comprise those from P3 onwards.

Assuming the presence of a separate gilt pool, suggests an extra cost for the additional parity gestation accommodation of 2.5% and 1.4% for additional labour, assuming current labour cost is 13% of total CoP. Special diets for the gilts and young sows. Most progressive breeders should be feeding the gilt developer and gilt lactation diets, so there should be no extra cost. Discussions with American feed compounder indicated that these would be 16% more expensive than conventional diets. It was calculated that a proportion of a herd with a 5.5 litter average SPL Sow production life, such an dietary regime will raise CoP by 3.0%.

Interviews suggest that this depends whether the breeder has spare and separate nursery rooms dedicated to these animals and not shared or transferred to others, even after a thorough all-in/all-out cleaning process.

Perhaps the best advice for the time being is to understand the mechanisms of the concept and do all you can do to minimise contact between the sows in the first two parities from those in subsequent ones and likewise for their offspring in the nurseries. There are still unanswered questions. The costs of converting an existing breeding farm into two separate sections have not been fully explored and recorded. However, anyone considering building a new unit, or extensively re-modelling an existing one, should think carefully about a separate section of the farm for the first two parities. For a really big organization, two or more separate farms. This could be the way things are going to move in the future.

3. New Words and Phrases

exclusively [ɪk'sklusɪvlɪ] *adv.* 唯一地；特定地

vague [veɪg] *adj.* 模糊的；(思想上) 不清楚的；(表达) 含糊的

entail [ɪn'teɪl] *v.* 需要；牵涉；使必要；限定继承

assumed [ə 'su:md] *v.* 假定；假设；取得（权力）assume的过去式和过去分词；呈现

extend [ɪk'stend] *v.* 延伸；扩大；推广

strain [stren] *n.* 血统，家族

gestation [dʒe'steɪʃn] *n.* 妊娠期

comprise [kəm'praɪz] *vt.* 包括；由……组成

presence ['prezns] *n.* 出席；仪表；风度

accommodation [əˌkɑːmə'deɪʃn] *n.* 住处；适应；和解；便利

labour ['lebər] *n.* 分娩

conventional [kən'vɛnʃənəl] *adj.* 传统的；平常的；依照惯例的

proportion [prə'pɔ:rʃn] *n.* 比例

dietary ['daɪɪˌtɛri] *adj.* 饮食的，规定食物的

regime [reɪ'ʒi:m] *n.* 管理，方法；（病人等的）生活规则

contact ['kɑ:ntækt] *v.* 联系；接触

subsequent ['sʌbsɪkwənt] *adj.* 随后的；后来的

likewise ['laɪkˌwaɪz] *adv.* 同样地；也，而且

convert [kən'vɜ:rt] *v.* 转变；转换

section ['sekʃn] *n.* 节；部门；部分；部件

explore [ɪks'plɔ:d] *v.* 勘查；探索，探究，仔细查看

4. Notes to the Text A

(1) Additionally there is evidence that Parity Segregation is much more likely to extend the sow productive life of the herd (SPL).

分胎饲养很可能会延长母猪生产利用年限的依据是，分胎饲养提供了更好地照顾，使得母猪的免疫系统得到了更好的发育，减少了它们在第一胎和第二胎时受到的伤害，从而增加了母猪的使用年限。

(2) Assuming the presence of a separate gilt pool, suggests an extra cost for the additional parity gestation accommodation of 2.5% and 1.4% for additional labour, assuming current labour cost is 13% of total CoP.

后备母猪、初产母猪的免疫系统和每日营养需求与经产母猪不同，因此可以对其建设独立的饲养区。但是这会导致额外的成本增加，这个比例大概使建筑成本增加2.5%，人工成本增加1.4%。如果母猪群平均生产使用寿命是5.5胎，这种饲喂模式会增加3.0%的运营成本。

(3) Perhaps the best advice for the time being is to understand the mechanisms

of the concept and do all you can to minimise contact between the sows in the first two parities from those in subsequent onesand likewise for their offspring in the nurseries.

要彻底做到分胎饲养，还需要为小猪提供独立的保育舍，并保证即使在进行彻底的全进全出清洗过程后也不会与其他小猪共享或转移给其他小猪。然而，这样做的成本过高，因此我们眼下能做到的是先掌握此模式的机理，并尽量隔离开初产母猪、经产母猪以及保育猪，这样做也许能对生产提供帮助。

5. Exercises

(1) Fill in the blanks by finishing the sentences according to the passage

①This development should be of particular interest to pig breeders due to growing evidence of its influence on ________ .

②Because the gilt and possibly the young P1 sow, and even the P2 sow, are such a source of infection due to their ________ , medication costs are reduced in the more mature sows and their offspring by up to ________ , and a 20% reduction looks to be typical.

③ Big difference in micro - mineral intakes between the first and second pregnancies and those of the more ________ .

④It was calculated that as a proportion of a herd with a 5. 5 litter average SPL, such an dietary regime will raise CoP by ________ .

⑤Interviews suggest that this depends whether the breeder has ________ and ________ nursery rooms dedicated to these animals and not shared or transferred to others, even after a thorough all-in/all-out cleaning process.

⑥Perhaps the best advice for the time being is to understand the mechanisms of the ________ and do all you can to minimise contact between the sows in the first ________ from those in subsequent ones and likewise for their offspring in the nurseries.

(2) Answer the following questions according to the passage

①How do you think of parity segregation before reading this article?

②When did the sow will have a derdeveloped immune systems?

③Comparing with the gilt and possibly the young P1 sow, and even the P2 sow, how many medication costs can be reduced in the more mature sows and their offspring?

④How many extra cost for the additional parity gestation accommodation and additional labour?

⑤What can we do for parity segregation at the moment?

⑥Do you think parity segregation will be achieved in the future? Why?

(3) Translation of the following sentences into Chinese

①Pig breeders are vague about what it entails and have assumed that it means raising sows separately, parity by parity.

②Additionally there is evidence that Parity Segregation is much more likely to extend the sow productive life of the herd (SPL).

③Big difference in micro-mineral intakes between the first two pregnancies and those of the more mature sows.

④If this is typical surely we can no longer afford to feed the same gestation diets to at least the P0 and P1 sows (and probably the P2 sows as well) compared to the rest of the herd which comprise those from P3 onwards.

⑤The costs of converting an existing breeding farm into two separate sections have not been fully explored and recorded.

⑥However, anyone considering building a new unit, or extensively re-modelling an existing one, should think carefully about a separate section of the farm for the first two parities.

课文A　胎次分段

鉴于越来越多的证据表明，养猪生产者对分胎饲养特别关注是因为母猪分胎次进行饲养会对疾病的发生造成影响。尽管母猪的分胎饲养模式在学术界已经讨论了至少5年，但仅在过去2年，人们才开始关注这一概念，并且几乎全部都是通过出自美国的研究论文和技术会议来了解。在其他地方，养猪生产者对这种模式的详情很不了解，并且误认为是按胎次把母猪分开饲养。

由于后备母猪、第1胎母猪甚至第2胎母猪的免疫系统尚未发育成熟，它们往往成了疾病的感染源，相比之下免疫系统发育较为完善的母猪及它们的后代，用药成本最多可降低50%，通常情况下也可降低20%。另外，有证据表明，分胎饲养很可能会延长母猪生产利用年限。

与分胎饲养模式有关的一个未来研究领域可能是，第1、第2胎母猪，甚至是第3胎母猪（如果来自高产母猪品系）所产的后代，与由免疫系统更为成熟的第4胎至第7、8、9、10胎母猪所产的后代相比，需要补充获得各种均衡的营养和/或营养含量。母猪在前2个妊娠期与在免疫系统更为成熟的高胎次妊娠期相比，微量元素需求量上的差异很大。如果这种情况确凿，那么我们再也不能向第0胎和第1胎甚至第2胎的母猪提供与第3胎起的高胎次母猪相同的妊娠期日粮了。

如果建设独立的后备母猪饲养区，额外增加的妊娠舍建设成本和饲养人员成本分别为2.5 %和1.4 %（假设一般劳动成本占总运营成本的13 %）。后备母猪和低胎次母猪还需要专用日粮。目前大多数先进的生产者都已在使用小母猪生长期和哺乳期日粮，因此这方面不会再产生额外成本。与美国饲料生产商讨论后表明，专用日粮在价格上要比传统日粮高出16 %。经计算发现，如果母猪群平均生产使用寿命是5.5胎，这种饲喂模式会增加3.0%的运营成本。

调查表明，这取决于养猪场是否有这些猪专用且独立的保育舍，即使在进行彻底的全进全出清洗过程后也不会与其他小猪共享或转移给其他小猪。

或许眼下最合适的建议是掌握此模式的机理，并竭尽所能的减少前2胎母猪与经产（3胎及以上）母猪的接触，以及与保育猪之间的接触。目前仍有很多问题悬而未决。将现有养猪场分解成两个独立区域所需的成本还没有得到充分的探讨和分析。然而，考虑新建或者大规模改建现有猪舍的养猪生产者，应该仔细考虑为第一、二胎次的小母猪在养猪场中建立一个属于它们的独立区域。对一家真正的养猪企业而言，建立两个或更多独立的养猪场可能是将来要考虑的生产模式。

Part Two Extensive Reading

1. Text B Not Spotting the Cause of Slow Growth Promptly is Costly

Failure to detect and act on 10% reduction in daily gain in pigs of 60 kg across a 4 week period results in 15 kg less MTF (meat produced per tonne of feed) *e. g.* 15 kg less saleable meat for every tonne fed then and thereafter. This is equivalent to a CNY 158. 68 per tonne increase in the cost of grower's food, or about a 12% price rise. Add typical overhead costs to this and the 12% becomes 18%—on some farms 22%.

Even 5% shortfall in growth (say around 40 g/day or a quarter of a kg/week) on what is feasible today is barely noticed by the producer—but it is equivalent to having to pay CNY 39. 67 per tonne more for his growing finishing food.

What cause poor growth rate?

The causes of these poor action level growth rates will be found to be are as follows.

(1) A larger than acceptable post weaning check.

(2) Lack-lustre growth once the post weaning check is surmounted. This is usually due to disease, with respiratory infection the most common cause. As well as veterinary consultation, an audit of the ventilation system especially in winter is advisable.

(3) The 12~16 week check. The reasons for this phenomenon are unclear but the following checks are advised. Those producers recording weekly growth rates do see it on their graphs while it may go unnoticed otherwise.

(4) After 16 weeks to slaughter the pig's immune status, peck-order, appetite and thermoregulatory system should be well-established. The areas to check here are disease, again respiratory disease, ileitis/colitis and overcrowding. Too stuffy air rather than a too cold environment is usually the culprit, both in summer and winter. Occasionally nutrition is at fault, in my experience poor amino acid balance can be responsible, but also overfeeding protein in the last month before market weight is rather too common. This protein would have been better fed in the first third of the pig's life.

(5) Temperature is often at fault.

2. Notes to the Text B

Failure to detect and act on 10% reduction in daily gain in pigs of 60 kg across 4 weeks period results in 15 kg less MTF (meat produced per tonne of feed) *e.g.* 15 kg less saleable meat for every tonne fed then and thereafter.

Failure to do 未做成。

3. Answer Questions

(1) How many kg less MTF do failure to detect and act on 10% reduction in daily gain in pigs of 60 kg across 4 weeks period results in?

(2) What causes the poor growth rate of growing pigs?

(3) Would protein have been better fed in the first third of the pig's life?

课文B 不能及时指出猪只增长缓慢的原因会造成巨大损失

假设体重为60kg的猪只平均日增重降低10%，若此原因未被及时发现并采取相应措施，而持续影响4周时间，那将导致每吨饲料的产肉量降低15kg。这相当于每吨生长猪饲料的成本增加158.68元人民币，饲料成本大约增加12%。若考虑经营管理费用的增加，所增加的成本将由12%变成18%，有些猪场甚至可高达22%。

在当今的猪只生产条件下，即使猪只生长速度降低5%（即40g/d或250g/周）也很难被生产者所发现。但其所带来的影响相当于每吨生长育肥猪饲料成本增加39.67元人民币。

什么原因造成猪生长缓慢？

导致猪只生长速度缓慢的原因如下所示。

（1）仔猪受断奶后生长滞缓的影响超过了可接受的程度；

（2）断奶仔猪生长滞缓期过后生长出现缓慢。该状况多因疾病所导致，呼吸道疾病是最主要的影响因素。建议及时咨询兽医，特别在冬天时应对猪舍通风状况进行检查。

（3）12~16周龄间猪只生长滞缓。导致猪只生长滞缓的原因尚不清楚，建议进行生产速度的测定与记录，每周测量并记录猪只生长速度的生产者可从猪群生长曲线图中发现此现象，否则，此阶段的生长缓慢常被忽视。

（4）从16周龄至上市期间，猪群的免疫状态、社会层级关系、食欲及温度调控系统逐步发育完善。此阶段造成生长缓慢的主要因素是疾病，如呼吸系统疾病，回肠炎或结肠炎以及猪群密度过大。无论在夏天还是冬天，猪舍内空气不流通往往比温度低对猪群的影响更大。有时也可能因饲料配方错误所导致。依笔者经验看，氨基酸不平衡是其可能的原因，但在猪只达到上市体重前1个月饲喂过量的蛋白质也是所导致生长缓慢的常见原因。而这些蛋白质在猪只生长周期的前1/3的时间内添加效果会更好。

（5）猪舍温度设置不当。

扫码进行拓展学习

Lesson 17

Poultry

Part One Intensive Reading

1. Learning Objective

After learning this lesson, you should understand the following:

(1) Describe what GI is and its role in poultry.

(2) Give the reason why a mature gut flora is slow to develop under the condition of modern large-scale poultry production.

(3) Give an overview of whom and what the high susceptibility of the young bird and its lack of competitive gut flora were highlighted by.

(4) Describe Nurmi's theory.

(5) Define competitive exclusion and the CE concept involves.

2. Text A Competitive Exclusion

The gastrointestinal (GI) tract of an adult bird harbors complex populations of microbes, and these organisms play an important role in maintaining the health and well-being of the host. The mature gut flora competes effectively with any invading organisms that may be harmful to avian or human health and can prevent them from colonizing the digestive tract. Microbial colonization of the GI tract normally begins soon after hatching and especially when the bird starts eating. However, under the conditions of modern, large-scale poultry production, a mature gut flora is slow to develop. This is because the birds are hatched and reared initially in a highly sanitized environment, and there is no contact with the mother hen. In the 1950s it had been observed that the resistance of young chicks to *Salmonella* colonization increases with age, so that 2-week-old birds were more difficult to infect, even with relatively high doses of *Salmonella*, but there was no explanation of the phenomenon at that time.

The high susceptibility of the young bird and its lack of competitive gut flora were highlighted by Esko Nurmi in Finland 35 years ago. In 1971, the Finnish broiler industry suffered from a widespread outbreak of *Salmonella* infants infection, the origin being a contaminated lot of raw feed material. The majority of broiler flocks became *Salmonella*—positive, and at the same time the incidence of human cases from this serotype increased considerably. In an attempt to solve the problem of *Salmonella* infection in poultry, research was begun at the National Veterinary and Food Research Institute. Nurmi and his research group were the first to demonstrate experimentally that administering intestinal contents from healthy adult birds to newly hatched chicks prevented them from becoming colonized by *Salmonella* and that the protective microflora could be cultured for administration by a relatively simple method. The Nurmi approach has been widely adopted in different countries in relation to *Salmonella* control in poultry and is referred to as the Nurmi concept or competitive exclusion (CE). The treatment has been defined as "the early establishment of an adult-type intestinal microflora to prevent subsequent colonization by enteropathogens."

CE treatment was originally intended to prevent intestinal colonization of young chicks with food-poisoning *salmonella*s, but over time the approach has been extended to cover other human and poultry enteropathogens (*e.g.*, pathogenic strains of *Escherichia coli*, *Campylobacter* spp., *Clostridium perfringens*, and *Listeria monocytogenes*). Improvements in bird performance have also been shown in well-controlled laboratory-scale studies as well as in the field.

In this unit, the development and applicability of the CE concept in poultry management, and factors affecting the efficacy of the treatment are reviewed and discussed. In addition, the nature of the protective mechanism and differences observed between various CE preparations and other types of probiotics are considered.

The competitive exclusion of one type of bacterium by others was a term first used by Greenberg. He studied the intestinal flora of blow-fly maggots and claimed that exclusion of *S. typhimurium* from the maggots was so effective that the organism survived in the gut only if the normal microflora was simplified or eliminated. A similar phenomenon in higher animals had been demonstrated earlier by Luckey. Colonization resistance, a term synonymous with CE, was introduced by van der Waaij et al. when examining the intestinal flora of mice. The term competitive exclusion was used in relation to poultry for the first time by Lloyd et al.

The CE concept involves the following points, as stated by Pivnick and Nurmi:

①Newly hatched chicks can be infected by only a single cell of *Salmonella*.

②Older birds are resistant to infection because of the autochthonous microbiota of the gut, particularly in the ceca and colon, but possibly in other parts of the GI tract as well.

③Chicks hatched by sitting hens are probably populated more rapidly by the autochthonous gut microflora of the adult.

④Hatcheries have replaced sitting hens, and the mass production of chicks is carried out in such a sanitary environment that the autochthonous microfiora is not introduced at the hatching stage.

⑤The growing houses in which newly hatched chicks are placed are usually thoroughly sanitized and the floors covered with fresh litter before new batches of birds arrive. Thus, the autochthonous flora of the adult is not readily available to populate the gut of the chicks.

⑥The artificial introduction of an adult intestinal microflora makes most of the recipient chicks immediately resistant to $10^3 \sim 10^6$ infectious doses of *Salmonella*.

⑦The intestinal flora of adult birds can be introduced as a suspension of fecal or cecal material, or as an anaerobic culture of such material. The treatment preparation may be introduced directly into the crop or by addition to the drinking water and possibly the feed. Spray treatment is also possible.

⑧The source of the treatment is usually the homologous bird species, although preparations derived from chickens will protect turkeys, and vice versa.

3. New words and phrases

gastrointestinal tract 消化道

harbor [ˈharbər] *n.* 港口；*v.* 隐藏，庇护，藏匿

microbe [ˈmaɪkrəʊb] *n.* 微生物，细菌

organism [ˈɔːgənɪzəm] *n.* 有机体；生物体；微生物；有机体系，有机组织

gut flora 肠道菌群；菌丛

avian [ˈeɪviən] *adj.* 鸟的，鸟类的

colonize [ˈkɒlənaɪz] *vt.* 将……开拓为殖民地；从他地非法把选民移入；移于殖民地；[生] 移植植物 *vi.* 开拓殖民地；移居于殖民地

invade [ɪnˈveɪd] *vt.* & *vi.* 侵入，侵略；进行侵略；蜂拥而入，挤满；（疾病，声音等）袭来，侵袭 *vt.* 侵犯；侵袭；涌入；干扰

sanitize [ˈsænɪtaɪz] *vt.* 净化；进行消毒；使清洁；审查

Salmonella [ˌsælmə'nelə] *n.* 沙门菌

serotype ['sɪərətaɪp] *n.* 血清型 *v.* 按血清型分类，决定……的血清型，把……按血清型分类

flock [flɒk] *n.* 兽群，鸟群；群众；棉束；大堆，大量 *vi.* 群集，成群结队而行 *vt.* 用棉束填

enteropathogen 肠病原体

pathogenic *adj.* 引起疾病的

Escherichia Coli 大肠杆菌

Clostridiumperfringens [klɑs'trɪdɪəm'pɜːfrɪndʒenz] 产气荚膜梭菌

Listeria monocytogenes 单核细胞增多性李斯特菌

probiotics *n.* 益生菌

preparation *n.* 制剂，配制品

maggot ['mægət] *n.* 蛆；*adj.* 多蛆的

S. typhimurium 鼠伤寒沙门菌

synonymous [sɪ'nɒnɪməs] *adj.* 同义词的；同义的，类义的

blow-fly 绿头苍蝇

autochthonous [ɔː'tɑːkθənəs] *adj.* 当地的；土生的；独立的；地方性的

microbiota [maɪkrəʊba'ɪɒtə] 微生物丛，微生物区

cecal ['siːkl] *adj.* 盲肠的

colon ['kəʊlən] *n.* 结肠

populate ['pɒpjuleɪt] *vt.* 居住于；生活于；移民于；落户于

hatchery ['hætʃəri] *n.*（尤指鱼的）孵化场

litter ['lɪtər] *n.* 杂物，垃圾；（一窝）幼崽；褥草；轿，担架 *vt.* & *vi.* 乱扔；使杂乱；乱丢杂物；使饱含

recipient [rɪ'sɪpiənt] *n.* 接受者；容器；容纳者；*adj.* 容易接受的；感受性强的

suspension [sə'spenʃn] *n.* 悬浮；悬架；悬浮液；暂停

homologous [hə'mɒləgəs] *adj.* 同源的；相应的；类似的；一致的

vice versa 反之亦然

4. Notes to the Text A

(1) The gastrointestinal (GI) tract of an adult bird harbors complex populations of microbes, and these organisms play an important role in maintaining the health and well-being of the host.

句中“play an important role in…”指“在某方面起着很重要的作用”。

（2）文章出现很多像“Barrow et al.”，“Nisbet et al.”人名后加“et al.”，表示“其他人、等人”。

（3）In the 1950s, it had been observed that the resistance of young chicks to *Salmonella* colonization increases with age, so that 2-week-old birds were more difficult to infect, even with relatively high doses of *Salmonella*, but there was no explanation of the phenomenon at that time.

句中“it had been observed that…”引导主语从句，“it”是形式主语，真正的主语是“that”后的从句；后面的“so that…”并不是我们常见的“so that”句型，所以不能译为“为了，以便”，这里翻译为“所以”。

（4）Nurmi and his research group were the first to demonstrate experimentally that administering intestinal contents from healthy adult birds to newly hatched chicks prevented them from becoming colonized by *Salmonella* and that the protective microflora could be cultured for administration by a relatively simple method.

此句中“the +序数词+to do sth.”是一个英语中的常见句型；“demonstrate that+宾语从句”，从句的主语较长“administering intestinal contents from healthy adult birds to newly hatched chicks”和“the protective microflora”，谓语动词是“prevent… from+doing sth.”和“could be cultured.”

（5）In addition, the nature of the protective mechanism and differences observed between various CE preparations and other types of probiotics are considered.

句中“in addition”表示“除此之外”；“observed”是过去分词做定语，修饰前面的名词，真正的谓语动词是“are considered”。

（6）He studied the intestinal flora of blow-fly maggots and claimed that exclusion of *S. typhimurium* from the maggots was so effective that the organism survived in the gut only if the normal microflora was simplified or eliminated.

句中“so…that…”引导结果状语从句，译为“如此……以至于……”；“only if”表示“只有……才；只有在……的时候，唯一的条件是……”；而“if only”是很容易和“only if”混淆的句型，常常用来表达强烈的愿望或遗憾，因此，主要用在虚拟语气中，常被译为“但愿”，如“If only she could be my sister.”译为“她要是我的妹妹该多好啊！”说明她不是我的妹妹。

5. Exercises

（1）Fill in the blanks by finishing the sentences according to the passage.

①The mature gut flora competes effectively ________ any invading organisms that may be harmful to ________ or human health and can prevent them from

________ the digestive tract.

②However, under the conditions of modern, large-scale poultry production, a mature gut flora is ________ to develop.

③In 1971, the Finnish broiler industry ________ from a widespread outbreak of *Salmonella* infantis infection, the origin being a ________ lot of raw feed material.

④Improvements in bird performance have also been shown in well-controlled laboratory-scale studies ________ in the field.

⑤The competitive exclusion of one type of bacterium by others was a ________ first used by Greenberg.

(2) Answer the following questions according to the passage.

①What role does the mature gut flora play?

②Why is a maturegut flora slow to develop under the conditions of modern, large-scale poultry production?

③When and who did highlight the high susceptibility of the young bird and its lack of competitive gut flora?

④What is Nurmi's theory?

⑤What does CE mean?

(3) Translate the following sentences into English.

①This is because the birds are hatched and reared initially in a highly sanitized environment, and there is no contact with the mother hen.

②In an attempt to solve the problem of *Salmonella* infection in poultry, research was begun at the National Veterinary and Food Research Institute.

③The Nurmi approach has been widely adopted in different countries in relation to *Salmonella* control in poultry and is referred to as the Nurmi concept or competitive exclusion.

④The term competitive exclusion was used in relation to poultry for the first time by Lloyd et al.

⑤The source of the treatment is usually the homologous bird species, although preparations derived from chickens will protect turkeys, and vice versa.

课文 A　家禽管理中的竞争性排斥处理

一个成年家禽的胃肠道中含有大量的复杂微生物，这些有机生物对寄主的健康起着非常重要的作用，成熟的菌群能有效地与外来入侵的致病菌进行对抗，阻止它们在消化道中占有优势，胃肠道的微生物占领从孵化之后就开始了，特别是在家禽开始进食的时候，可是，在现代的大规模集约化生产条件下，家禽的成熟肠道菌落形成很慢，这是因为禽类起初在高度消毒的条件下孵化和饲养，并且和母鸡没有联系。在20世纪50年代，我们发现雏鸡对沙门菌的抵抗能力随着年龄增加而增加，2周龄的鸡即使在沙门菌高剂量的条件下也难感染，但是在那个时期并没有对这个现象进行解释。

芬兰的Esko Nurmi在35年前特别强调了雏鸡的高度易感性和占有优势的肠道菌落数量的缺乏。在1971年，芬兰肉鸡工业遭受了一个雏鸡沙门菌的大面积爆发，起因就是原料被污染，鸡群中的大多数鸡只出现沙门菌阳性，同时人类发生感染的概率也大大地增加，国家兽医和食品研究院开始进行研究，尝试着要解决禽类沙门菌感染的问题。Nurmi和他的研究课题组是第一个用试验证明这个理论的。它们把健康成年家禽肠道的微生物菌落接种到刚孵化的雏鸡肠道内，以免它们肠道内沙门菌数量占优势，具有保护作用的菌落可以用相当容易的方法培养。Nurmi的研究已经在沙门菌爆发的国家得到广泛认可，被称为Nurmi概念或竞争排斥（E）。这个治疗被定义为“为了阻止肠道病原体占有优势对成年家禽肠道微生物菌落的早期建立”。

CE治疗目的起初是阻止幼禽在进食了饲料中含有有毒的沙门菌使其在肠道内大量繁殖，但是随着时间的推移，这个方法已经延伸至揭示别的人类和家禽的致病菌（如大肠杆菌的致病菌株、弯曲杆菌属、产气荚膜梭菌和单核细胞增多性李斯特菌），我们可以在控制良好的实验室以及生产中看到禽类生产性能提高了。

本单元讨论了CE的概念在禽类管理中的应用和发展以及影响治疗有效性的因素，除此以外，保护机制的机理以及各种各样CE制剂及其他种类的益生菌所观察到的差别都被提及。

Greenberg首先使用一种细菌被另一种细菌竞争排斥的术语。他研究了蝇蛆的肠道菌群，声称蝇蛆的肠道菌群能非常有效地将鼠伤寒杆菌排除在蝇蛆之外，只有当正常菌群被减少或消除时，这种微生物（鼠伤寒杆菌）才能在肠道中存活。类似的现象在高等动物身上已经被luckey证实了。定殖抵抗，是CE的同义词，是由van der waaij等人在检测小鼠肠道菌群时提出的。竞争性排斥这一术语首次被Lloyd等用于家禽中。

CE 的概念包括以下几点，如 Pivnick 和 Nurmi 所述：

①仅单个细胞的沙门菌就能感染刚孵出的雏鸡；

②较大的鸟类对感染有抵抗力，因为它们的肠道中存在自身微生物群，特别是在盲肠和结肠中存在自身微生物群，但在消化道的其他部分也可能存在；

③母鸡孵化的雏鸡可能更快地受到成鸡体内来源的肠道微生物菌群的影响；

④孵化场取代了母鸡的孵化方式后，大规模生产雏鸡环境的卫生好，原生微生物菌群孵化阶段没有在雏鸡体内建立；

⑤在新一批雏鸡到来之前，饲养新出壳雏鸟的育雏室，通常要进行彻底消毒，地板上要铺上新鲜的垫料，因此，成年鸡携带的原生菌群无法随时进入雏鸟的肠道；

⑥人工接种成鸡肠道微生物菌群，可使大多数受体鸡立即产生对 10^3~10^6 剂量的沙门菌感染的抵抗力；

⑦可以用成年鸡的粪便或盲肠物质的悬浮物作为肠道菌群接种，或者作为该物质的厌氧培养物接种。处理制剂可以直接在生产中接种，或者通过添加在水中饮用和料中饲喂，亦可喷雾处理；

⑧虽然从鸡身上提取的制剂可以保护火鸡，从火鸡身上提取的制剂也可以保护鸡。但通常治剂的来源要是同源的禽类。

Part Two Extensive Reading

1. Text B Feeding and Management of Young Chickens

The object to be accomplished in the feeding of young chickens is to grow strong, well-matured, normal individuals with little or no mortality, at a cost that is consistent with good results.

Changesin weight during growth

One of the purposes of a chick ration is to produce normal growth. Normal growth refers to the weights, limited by inheritance, which the chicks should attain when kept under favorable conditions. It is necessary to keep in mind, however, that the rate of growth is subject to variation, depending upon breeding, management, and environmental conditions. Hence there are definite standards for each variety or strain which has a different weight at maturity.

The actual weights of chicks, and especially the weights during the early growth period, have improved owing to advances in breeding, feeding, and management. Weights and feed consumption representative of earlier results.

There is very little difference in the growth of chickens that are reared indoors or on range when they are fed equivalent rations. However, Kentucky reported that the pullets and cockerels reared indoors and without direct sunshine were not so vigorous and active or so healthy in appearance as those reared outdoors.

The time of hatching does have an influence upon growth. The general tendency is for early hatched chicks to grow faster than late-hatched chicks. However, Kempster also reports that these differences may be explained on the basis of high temperatures during the growing season. The type of growth curve, therefore, varies, depending on the date the chicks are hatched and the temperatures that prevail during the growing season.

The energy of the young stock is thrown into growth and motion. If either growth or activity is stopped for any period of time, the results will not be favorable. Both must continue steadily according to the laws of growth. Growth represents increases in tissues, such as bones, muscle, skin, feathers, and nerves. Besides water, these increases are chiefly in protein and minerals.

2. Notes to the Text B

(1) It is necessary to keep in mind, however, that the rate of growth is subject to variation, depending upon breeding, management, and environmental conditions.

句中“It is necessary to…”句型，“it”是形式主义，后面的动词不定式是真正的主语。此句“however”是插入语，“that”引导宾语从句。“depending upon”中动词 ing 形式做状语。

(2) Both must continue steadily according to the laws of growth.

句中“both”代词做主语，指“活动和生长”。

3. Answer questions

(1) What's the objective in the feeding of young chickens?

(2) What's the purpose of a chick ration?

(3) What's the rate of growth subject to?

(4) How to improve the actual weights of chicks especially during the early

growth period?

(5) Does the time of hatching have an influence upon growth?

(6) What does growth represent?

课文 B 雏鸡的饲养管理

在雏鸡极低的死亡率下，使其健壮生长，发育良好是我们在饲养管理过程中最重要的目标，为了完成这一目标，我们需要耗费大量的人力和物力。

在生产过程中体重的变化

给雏鸡提供饲料的目的之一就是使其正常生长发育。正常的生长发育指的是体重尽管受遗传限制，但在生长环境有利的情况下仍能体重达标。但必须要谨记生长率容易受育种水平、管理水平和环境条件的影响。然而对于不同的成熟雏鸡品系有着不同的体重标准。

雏鸡的实际体重，特别是在早期生长阶段的体重由于育种水平、饲养管理水平的提高而出现改善。生长早期的结果显示，体重和饲料消耗的代表性数据在最近更新的数据有了更大的提高。

室内圈养的雏鸡与散养的相比，在同样的饲料配给情况下，生长率几乎没有差别。可是，Kentucky 的报道中指出，仅从外表看，在室内圈养不受直接阳光照射的小母鸡和小公鸡不如在户外散养的有活力和健康。

孵化的时间对生长有一定影响。一般的趋势是早孵化出的小鸡比晚孵化出的小鸡生长速度更快。可是，Kempster 也指出这些差别可以以生长季节温度的理论基础来解释。因此，生长曲线的变化取决于小鸡孵化的日期和生长的季节中占主导的气温。

幼禽的能量主要用于提供生长和运动。如果在任何时期它们停止生长或运动，那么将导致一个不利的结果。所以必须要按照生长规律稳定地生长和运动。生长表现为组织器官的生长发育，比如骨骼、肌肉、皮肤、羽毛和神经。除了水以外，这些组织器官的生长主要体现在蛋白质和矿物质的含量上。

扫码进行拓展学习

Lesson 18

Sheep

Part One Intensive Reading

1. Learning Objective

After learning this lesson, you should understand the following:

(1) Explain why the lamb is best fed on Mama, but at times you need to help get the process going.

(2) Let the lamb nurse by itself if it will, but do not let more than half an hour to an hour pass without its nursing.

(3) Explain why it is very important to make sure every lamb can get enough colostrum.

(4) Give an overview of how to decide whether the lamb is actually getting any milk.

(5) Describe how to help the lambs find the teats and grab teats.

(6) Describe how to vaccinate the ewe with a vaccine to protect against tetanus, enterotoxemia, and other common clostridial diseases.

(7) Describe what can do if an ewe has too much milk.

2. Text A Feeding the Lamb

The lamb is best fed on mama, but at times you need to help get the process going. When the ewe stands up, she'll nudge the lamb toward her udder with her nose. The lamb is born with the instinct to look for her teats and is drawn to the smell of the waxy secretion of the mammary pouch gland in mama's groin. If the udder or teats are dirty with mud or manure, a swab with a weak chlorine bleach solution before the lamb nurses will clean things up and help prevent intestinal infection in the lamb.

Let the lamb nurse by itself if it will, but do not let more than half an hour to an hour pass without its nursing, as the colostrum (the ewe's first milk after lambing) provides not only warmth and energy but also antibodies to the common disease organisms in the sheep's environment. We usually opt not to interfere for about 20 minutes after the lamb is up on all fours and looking for a teat. If after 20 minutes it hasn't found the teat, is trying to nurse but doesn't seem to be getting any milk, or the mom won't let it nurse, we intercede. (This is easy with some ewes, which will stand really still until you and the lamb get things settled, but it can be a real pain with other ewes that just want to wander around in circles and carry on the whole time.) Occasionally, the ewe will not allow the lamb to nurse because she is nervous, has a tender or sensitive udder, or is rejecting the lamb. If the udder appears sensitive, it is often because it is tightly inflated with milk.

Remember the rule of thumb when you're trying to decide whether the lamb is actually getting any milk: a lamb that is getting milk will have its little tail whipping back and forth like a metronome at full speed. When a lamb is getting milk, its body fills out quickly, its skin folds start to disappear, and its little stomach becomes tight. When a lamb is a few hours old and is crying continuously or has a cold mouth, it is not nursing.

If the problem appears to be nothing more than a flighty ewe, but the lamb is still strong and trying to grab a teat, then restrain the ewe and allow the lamb to nurse. The ewe can be restrained with a head gate, with a halter, or by pushing her into a corner and leaning your weight against her.

If the lamb is getting on a teat but doesn't seem to be getting milk, you probably need to unplug the ewe. "Unplug" and means it literally: the end of the teat has been protected over the past several weeks by a little waxy plug, which is sometimes hard for a lamb to displace, especially if the lamb is a little weak to begin with. After you've broken the plug free, strip the teat of several squirts of colostrum by massaging down the teat between your thumb and index finger.

Sometimes lambs are a bit dumb and need you to help them find the teat in the first place. I've seen newborns try to nurse the front knee, wool on a tail, and other odd spots. Grab the lamb and force its mouth over the teat while you massage the teat to get a few squirts of milk in the lamb's mouth. Usually, once it gets those first couple of squirts, it settles down to business with no additional assistance. But occasionally you'll find a really slow one that you have to help for longer.

You may encounter a lazy lamb that, for no apparent reason, noes not want to nurse the ewe but will take a bottle with enthusiasm. These lambs can be maddeningly frustrating and can tax both your nerves and your patience with regard to how long you are willing to wait to see if they will begin nursing the ewes. We call these lambs "volunteer bummers"

If for some reason mom can't feed it any colostrum (no milk, bag hard with mastitis, too many previous lambs), then you'll need to feed it like a bummer lamb.

You have made a great contribution to the colostral protection of the new-born lamb if you have previously vaccinated the ewe (twice) with a vaccine to protect against tetanus, enterotoxemia, and the other common *clostridial* diseases. These antibodies are absorbed by the mammary gland from the ewe's bloodstream and are incorporated into the colostrum so they protect the new-born lamb until it starts to manufacture its own antibodies. The small intestine of the newborn lamb possesses a very temporary ability to absorb these large molecular antibodies from the colostrum. This ability to absorb antibodies decreases by the hour and becomes almost nonexistent by 16 to 18 hours of life. Colostrum is high in vitamins and protein and is a mild laxative, which can assist in passing the fetal dung (meconium, the black, tarry substance that is passed shortly after the lamb nurses for the first time).

The longer a lamb has to survive without colostrum, the fewer antibodies it has the opportunity to absorb and the less chance it has of survival if it develops problems. A weak lamb, or one of low birth weight, can be lost because of a delay in nursing.

When a ewe has too much milk, her udder becomes too full and the teats become enlarged. To rectify this situation, milk out a bit of this colostrum and freeze it in small containers for emergency use. Ice-cube containers and small resealable freezer bags are good options because they allow you to thaw small quantities as needed. Solidly frozen colostrum will keep for a year or more if it is well wrapped. When saving and freezing colostrum, you should have a combination of colostrum milked from several ewes, for they do not all produce the same broad spectrum of disease-fighting antibodies. Cow or goat colostrum can be stored and used in emergencies.

3. New Words and Phrases

lamb [læm] *n.* 羔羊；小羊

nudge [nʌdʒ] *vt.* 推进；(用肘) 轻推

udder [ˈʌdər] *n.* (牛、羊等的) 乳房

teat [tiːt] *n.* 乳头，乳头

waxy [ˈwæksi] *adj.* 像蜡的；蜡色的；苍白的

mammary [ˈmæməri] *n.* 乳房的，乳腺的

pouch [paʊtʃ] *n.* 小袋，育儿袋

groin [grɔɪn] *n.* 腹股沟，大腿根儿

manure [məˈnjʊər] *n.* 肥料，粪便；*v.* 给……施肥

swab [swɒb] *n.* （医用的）拭子，药签；*v.* 用拭子擦拭

chlorine [ˈklɔːriːn] *n.* 氯

bleach [bliːtʃ] *n.* 漂白，漂白剂

nurse [nɜːs] *n.* 喂，吃乳

antibody [ˈæntibɑːdi] *n.* 抗体

opt [ɒpt] *vi.* 选择，挑选

interfere [ˌɪntəˈfɪər]；*vi.* 干预，干涉；调停，排解；妨碍，干扰

intercede [ˌɪntəˈsiːd] *vi.* 说情；斡旋；调解，（为某人）说情

occasionally [əˈkeɪʒnəli] *adv.* 偶尔，偶然，有时候

tender [ˈtendər]；*adj.* 温柔的；嫩的

sensitive [ˈsensətɪv] *adj.* 敏感的；感觉的

inflate [ɪnˈfleɪt] *vt.* & *vi.* 使充气（于轮胎、气球等）；（使）膨胀

whip [wɪp] *vt.* 鞭打，抽打；严厉地折磨，责打或责备，迫使；把……打起泡沫

metronome [ˈmetrənəʊm] *n.* 节拍器

stomach [ˈstʌmək] *n.* 胃；腹部；食欲；欲望

continuously [kənˈtɪnjʊəslɪ] *adv.* 连续不断地，接连地；时时刻刻；连着；直

flighty [ˈflaɪti] *adj.* 反复无常的

literally [ˈlɪtərəli] *adv.* 逐字地；照字面地

squirt [skwɜːt] *n.* 喷，细的喷流；注射器

colostrum [kəˈlɒstrəm] *n.* 初乳

massage [ˈmæsɑːʒ] *n.* 按摩，推拿

assistance [əˈsɪstəns] *n.* 帮助，援助

apparent [əˈpærənt] *adj.* 易看见的，可看见的；显然的，明明白白的；貌似的，表面的；显见

enthusiasm [ɪnˈθjuːziæzəm] *n.* 热情，热忱；热衷的事物；宗教的狂热

frustrate [frʌˈstreɪt] *vt.* 挫败；阻挠；使受挫折 *adj.* 无益的，无效的

bummer [ˈbʌmər] *n.* 失望（或不愉快）的局面

tetanus [ˈtetənəs] *n.* 破伤风

enterotoxemia [ˌentərəʊtɒkˈsiːmɪə] *n.* （羊的）肠毒血病

clostridial [klɔsˈtridiəl] *adj.* 梭菌的，梭菌属的

incorporate [ɪnˈkɔːpəreɪt] *vi.* 合并；包含；吸收；混合

molecular [məˈlekjələr] *adj.* 分子的，由分子组成的

laxative [ˈlæksətɪv] *n.* 泻药；缓泻药

fetal [ˈfiːtl] *adj.* 胎儿的，胎的

meconium [məˈkəʊnɪəm] *n.* 胎尿，胎粪

tarry [ˈtæri] *vi.* 逗留；停留；暂住；徘徊 *adj.* 柏油的；涂柏油的，被柏油弄脏的

rectify [ˈrektɪfaɪ] *vt.* 改正，校正

thaw [θɔː] *vt.* 使融化

wrap [ræp] *vt.* 包；缠绕；用……包裹（或包扎，覆盖等）；掩护

spectrum [ˈspektrəm] *n.* 光谱；波谱；范围；系列

denature [diːˈneɪtʃə] *vt.* 使改变本性；使变质

4. Notes to the Text A

（1）The lamb is born with the instinct to look for her teats and is drawn to the smell of the waxy secretion of the mammary pouch gland in mama's groin.

句中"With+n/pron+to do sth"是英语中常见的独立主格结构，在句中做伴随状语，表示羔羊在出生后就有找乳头的本能。

（2）Let the lamb nurse by itself if it will, but do not let more than half an hour to an hour pass without its nursing, as the colostrum (the ewe's first milk after lambing) provides not only warmth and energy but also antibodies to the common disease organisms in the sheep's environment.

句中"as"引导原因状语从句；"not only… but also…"是个关联词组，用于连接两个表示并列关系的成分，着重强调后者，它的意思是"不仅……而且……"。

（3）This is easy with some ewes, which will stand really still until you and the lamb get things settled, but it can be a real pain with other ewes that just want to wander around in circles and carry on the whole time.

句中"which"在上句引导的是非限制性定语从句，先行词是"some ewes"，而后一句则是"that"引导的定语从句，先行词是"other ewes"。

（4）Remember my rule of thumb when you're trying to decide whether the lamb is actually getting any milk.

句中“a rule of thumb”是一个词组，意思为“一条经验法则”。

(5) The longer a lamb has to survive without colostrum, the fewer antibodies it has the opportunity to absorb and the less chance of survival it has if it develops problems.

句中“The +比较级，the+比较级”的句型是一个复合句，意为“越……越……”，其中前面的句子是状语从句，后面的句子是主句。

5. Exercises

(1) Fill in the blanks by finishing the sentences according to the passage.

①The lamb is best fed on ________, but at times you need to help get the process going.

②Let the lamb nurse by ________ if it will, but do not let more than ________ to ________ pass without its nursing, as the colostrum (the ewe's first milk after lambing) provides not only ________ and energy but also antibodies to the common ________ organisms in the sheep's environment.

③When a lamb is getting milk, its body ________ quickly, its skin ________ start to disappear, and its little stomach becomes ________. When a lamb is a few hours old and is ________ continuously or has a ________ mouth, it is not nursing.

④The ewe can be restrained with a ________ gate, with a ________, or by pushing her into a ________ and leaning your ________ against her.

⑤Grab the lamb and force its ________ over the teat while you ________ the teat to get a few squirts of milk in the lamb's ________.

(2) Answer the following questions according to the passage.

①How is the lamb is best fed? When will you give the lamb your help?

②How soon will you help the lamb nurse without its nursing?

③Why is colostrum important to the lamb?

④How to decide whether the lamb is actually getting any milk? How to help the lamb find the teats and grab teats?

⑤What vaccine have you previously vaccinated the ewe (twice)?

⑥How to keep the colostrum?

(3) Translate the following sentences into English.

①The lambis born with the instinct to look for her teats and is drawn to the smell of the waxy secretion of the mammary pouch gland in mama's groin.

②Let the lamb nurse by itself if it will, but do not let more than half an hour to an hour pass without its nursing, as the colostrum (the ewe's first milk after lambing) provides not only warmth and energy but also antibodies to the common disease organisms in the sheep's environment.

③Remember my rule of thumb when you're trying to decide whether the lamb is actually getting any milk: a lamb that is getting milk will have its little tail whipping back and forth like a metronome at full speed.

④The longer a lamb has to survive without colostrum, the fewer antibodies it has the opportunity to absorb and the less chance it has of survival if it develops problems.

⑤Loss due to lack of colostral antibodies is not the same as loss due to starvation, which occurs from receiving no milk at all.

课文A 羔羊饲喂

羔羊最好是由母羊喂养，但有时候需要你帮助母羊让这个过程继续下去，当母羊站立，她将用鼻子轻轻地把羔羊推向乳房。羔羊天生有寻找乳头的本能，被乳腺所分泌的乳白色液体的味道吸引到腹股沟。如果乳房或者乳头上有泥土或粪便，在羔羊哺乳之前，用蘸有低浓度的氯漂白液的棉签将其擦去，防止羔羊肠道感染。

如果羔羊吃乳，让它自己去吃乳，但是如果没有自己去吃乳，不要让时间超过30~60min，因为初乳（母羊产后第一次产的乳）不仅提供了能量和热量，还提供了对抗环境中常见的致病微生物的抗体。在羔羊四肢站立并正寻找乳头时，我们通常选择在20min以内不去影响羔羊的活动，而20min之后羔羊还没找到乳头，或者正在尽力吃乳但好像没有吃到乳，或者母羊不让它吃乳，我们需要干预了对于一些母羊来说很容易，它们会静静地站着，直到你和小羊把事情安顿好，但是对于那些在原地打转并且一直这样的母羊来说，干预是一种真正的痛苦，有时候母羊会因乳房敏感或者疼痛而焦躁，正在排斥羔羊，那么母羊将不会给羔羊哺乳，如果是乳房敏感，常因为是乳房中紧紧充满了乳汁。

当你正在努力判定羔羊是否吃到了乳，记录经验来看，正在吃乳的羔羊的尾巴会前后像一个节拍器一样有节奏地鞭打，当一头羔羊正在吃乳，它的身体会很快充胀，皮肤的皱褶开始消失，胃也变得胀起来。如果出生几小时的羔羊一直叫或者嘴巴很凉，那么说明这只羔羊没有吃到乳。

如果这个问题只是源自母羊的问题，而羔羊是强壮的，正在努力去捕捉乳头，那么限制母羊的活动可以让羔羊吃乳，可以用一个夹头部的闸门、笼头或者将母羊推入到一个角落，依靠自身的重量来限制她的活动。

如果一个羔羊正在吸吮乳头但是好像没有吃到乳，你可能需要除去乳头上的障碍物，去除障碍物的意思是：乳头末端在过去的好几周用蜡样的涂层保护着，这对羔羊来说是很难去除，特别是对于刚出生的弱羔。如果你已经去除涂层，通过你的拇指和食指向下按压乳头，有时候羔羊有点笨，需要你帮助它们找到乳头，我曾经看到新生羊羔把前膝、尾巴上的毛和身上其他的地方当作乳头努力吃乳，可以抓住羔羊，强迫它的嘴含住乳头，使乳汁进入羔羊的嘴里，通常，羔羊一感觉到乳汁就开始吸吮，不再需要外力帮助，但有时也会遇到慢性子的羔羊，需要多帮助一会儿。没有明显的原因，你可能遇到一只懒惰的羔羊，不想吃母羊的乳但是却非常喜欢用奶瓶喝乳，这类羔羊会产生令人发狂的挫败感，你愿意等着看需要多久它们才会吃母羊的乳，将会使你的神经和耐心承受很大的负担，我们把这类羊称为“天生的懒汉”。

如果因为某些原因母羊不能给羔羊哺喂初乳（如乳腺炎，之前有很多羊因此没有乳汁），那么你需要像喂无自食能力羔羊一样饲喂。如果你事先给母羊接种过疫苗去保护机体免受破伤风、肠毒血症和其他常见的梭状芽孢杆菌疾病，那么你就对新生羔羊的初乳保护做出了巨大贡献。这些抗体通过母羊血液被乳腺吸收，和初乳混合保护新生羔羊，直至羔羊建立自己的抗体，新生羔羊的小肠暂时有一个非常重要的能力就是能从初乳中吸收分子质量大的抗体，吸收抗体的能力按小时减退，在出生后的16~18h之后，这种能力几乎就不存在了。初乳中富含维生素和蛋白质，具有轻泻作用，有利于排出胎粪。

一只羔羊出生后获得初乳的时间越晚，它就会吸收更少的抗体，遇到问题就会有更少的存活机会。一头孱弱羔羊，或者出生体重低的羔羊可能会由于吃乳的延误而丧命。

一头母羊产生了大量的乳，她的乳房会充满，乳头将变大，为了改善这种情况，挤出一些初乳，将使其凝固在小的容器里以备紧急时所需，最好选择冰管的容器和小的可重复密封的容器，因为这样可以使你在需要的时候每次融化少量的初乳，凝固型初乳如果包裹好的话可以保存一年或者更久，在储存和凝固羊乳的时候，你应该把好几只羔羊的初乳混合起来，因为它们体内会产生不同的抗体，牛初乳和山羊的初乳应该储存起来以备不时之需。

Part Two Extensive Reading

1. Text B Sheep Care Guide

Because sheep are ruminants, they can utilize a wide variety of feedstuffs to meet their nutrient requirements. Extensively managed sheep operations typically use native forages, or improved pastures, and crop aftermath from grains, legumes, and vegetable crops. Intensively managed operations may use similar plant materials but may rely more heavily on harvested feeds including cereal grains. The nutritional needs of sheep in range flocks, farm flocks, and lamb feedlots vary greatly, and nutritional programs must be developed to address these specific, and sometimes unique, situations.

Basic nutrient groups include water, energy (carbohydrates and fats), protein, minerals, and vitamins. A sheep's nutrient requirements vary greatly and are heavily dependent upon such factors as age, sex, weight, body condition, stage of production, wool or hair cover, and environmental conditions, such as cold, wind, and mud. These factors are in a constant state of change, and the diet must be adjusted accordingly. The Nutrition Chapter of the Sheep Production Handbook provides sheep producers the necessary information to evaluate their sheep's nutritional needs in the various stages of condition and production and shows how to formulate diets that will adequately meet the sheep's needs for maintenance, growth, and reproduction. Additional information is available in the National Resource Council's publication, Nutrient Requirements of Sheep.

Forage quality may vary considerably throughout the year and from year-to-year. Routine monitoring of quality, using forage nutrient analyses, is very important to optimize animal efficiency, reduce costs, and maximize animal welfare. Periodic review of the nutrition program by a qualified nutritionist is advised. This is especially useful for producers who own small flocks and who have minimal experience formulating diets. Records of feeds fed and sources of feed ingredients are important to document the nutrition program and any feed additives used.

2. Notes to the Text B

(1) A sheep's nutrient requirements vary greatly and are heavily dependent upon such factors as age, sex, weight, body condition, stage of production, wool

or hair cover, and environmental conditions, such as cold, wind, and mud.

此句中有两个比较常见的词组,“be dependent upon⋯.”为“取决于⋯⋯”;“such⋯as.”为“诸如⋯⋯”;而最后句中出现的“such as”短语是用来解释前面提到的“environmental conditions”。

(2) Routine monitoring of quality, using forage nutrient analyses, is very important to optimize animal efficiency, reduce costs, and maximize animal welfare.

此句是前置性陈述,即在句中将主要信息尽量前置,通过主语传递主要信息,这在英语学术文章中非常常见。

(3) The Nutrition chapter of the Sheep Production Handbook provides sheep producers the necessary information to evaluate their sheep's nutritional needs in the various stages of condition and production and shows how to formulate diets that will adequately meet the sheep's needs for maintenance, growth, and reproduction.

句中“The Nutrition Chapter of the Sheep Production Handbook”为书名,属于专有名词,首字母大写;“how to formulate diets”为不定式短语,“diets”后的“that”从句是定语从句,先行词为“diets”。

3. Answer Questions

(1) What characteristics do ruminants have?

(2) Why do we need to develop nutritional programs for sheep?

(3) What do sheep nutrient requirements depend on?

(4) Why do we need to monitor routinely forage quality?

课文B 绵羊管理指南

因为绵羊是反刍动物,它们能利用许多种类的饲料来满足它们的营养需要,粗放型管理的绵羊,通常使用本地牧草或者改良牧草和谷类、豆类作物以及蔬菜作物,精细型饲养管理的绵羊可以使用相似的植物作物但是更多地依赖收割作物包括谷类作物,在某个固定区域放牧或者农场放牧绵羊以及羔羊的营养变化很大,必须制定出营养方案来记录这些有针对性的有时候甚至是独一无二的情况。

基本营养组成包括水、能量(碳水化合物和脂类)、蛋白质、矿物质和维生素。一头绵羊的营养需要变化很大,在很大程度上取决于年龄、性别、体重、身体状况、生产水平、毛发的覆盖面积以及环境状况,比如寒冷、风和泥土。这些因素处在一直变化的状态，那么日粮必须相应地调整。《绵羊生产手册》中营养章节给绵羊生产者提供了必要的信息去衡量不同环境阶段、不同生产阶段绵羊的营养需要，说明如何配制日粮能充分地满足绵羊的维持生存、生长和繁殖需要。在国内资源委员会的出版书中提供了绵羊的营养需要信息。

牧草的品质在不同年份、不同月份都有很大程度的变化。通过分析牧草营养成分来检测常规品质，对于动物利用率的最大化、经济成本最低和最大限度地提高动物福利是非常重要的。我们建议一个合格的营养师应该定期地检查营养配制方案，这对小型放牧的生产者以及配制经验非常少的生产者来说特别有用，饲料喂养记录和各种饲料原料来源、营养日粮配制方案和任何使用的饲料添加剂的记录都是非常重要的。

扫码进行拓展学习

Lesson 19

Lamb

Part One Intensive Reading

1. Learning Objective

After learning this lesson, you should understand the following:

(1) Colostrum provides fuel for heat productionand immunoglobulins for protection against infections.

(2) In order to meet its fuel needs, the newborn lamb requires large quantities of colostrum during the first 18 h after birth.

(3) The quantities needed during the first 18 h vary accordingly to the environmental conditions and the birth weight of the lamb.

(4) Giving oxytocin by injection, either intravenously or intramuscularly, just before milking the ewe greatly improves the yield of colostrum obtained.

(5) Using this method, milking ewes three times during the first day after birth will yield between 850 mL and 2400 mL of colostrum.

(6) Banking colostrum obtained from suitable donor ewes early in the lambing season is recommended.

(7) When deep frozen, it will keep without deteriorating for at least 1 year.

2. Text A Colostrums

It is well known that early intakes of colostrum by lambs during the first day after birth improve survival rates and it is obvious that measures which cater for orphans and lambs from ewes with insufficient colostrum would be beneficial. Neverthelesss, problems arise because farmers often do not know how much colostrum a lamb needs nor the quantities of colostrum which are available for fostered lambs or which are obtainable by milking; nor are they aware that large volumes of colostrum can be obtained from ewes quickly and effectively by hand milking.

Colostrum usually accumulates in the udder during the final few days of pregnancy. It is also produced during the first 12~24 h after birth, but is diluted progressively as milk production increase. The transition from colostrum to milk is gradual and is accompanied by marked decrease in the concentrations of antibodies and sodium, and by increase in potassium and lactose. The usual definition of colostrum emphasizes a high immunoglabium content, but in this unit the word colostrum will be used simply to mean early milk, *e. g.* all the exocrine secretions produced by the udder during the first 24 h after birth.

Lambs fed to appetite by bottle every 2 h can drink about 270 mL colostrum bodyweight (equivalent to 1080 mL for a 4 kg lamb) during the first 18 h after birth. This represents the maximum intake.

The amount of colostrum a lamb drinks depends on the quantities available and on the success of sucking. Colostrum availability is affected by breed, ewe nutrition during late pregnancy and the number of lambs born, whereas sucking success depends on the early establishment of a good ewe-lamb bond and on competition between litter mates. Obviously, colostrum intake can vary in different lambs between zero and the maximum. The amounts of colostrum a lamb needs depend mainly on how much fuel it requires for heat production. Therefore, any factor which increase heat production increases the colostrum requirement. During bad weather (cold, windy or rainy) the lamb must produce more heat to avoid hypothermia, its requirement for colostrum increases.

In order to avoid hypothermia, lambs born in average field conditions (0~10℃, windy, rainy) need about 210 mL colostrum/kg bodyweight during the first 18 h, while intakes of about 180 mL/kg would be adequate in housed animals (still, dry air at 2~10℃). These figures must be multiplied by the lamb's weight (kg) in order to estimate the total volume of colostrum required during the first 18 h after birth.

These quantities will normally also be sufficient to protect lambs against gut infections which cause watery mouth or diarrhoea because 200 mL of colostrum usually contain enough immunoglobulins for this purpose. Colostral immunoglobulins have specificities against particular pathogens and if that specificity is absent, protection will not be afforded. When lambs are to reared on commercially produced milk substitutes which contain no immunoglobulins, it is essential to provide them with at least 200 mL colostrum during the first day to reduce the risk of gut and other infections during early life.

These colostrum requirements represent the total intakes during the first 18 h after birth. When feeding orphan or sick lambs by stomach tube, it is important to avoid excessive stomach distension, so it is not wise to give more than 50 mL/kg on each occasion. It is convenient to use 50 mL syringes to deliver the colostrum as this saves times and permits the simple instruction that at each feed a lamb should receive one syringe-full for each kg of its weight. It is necessary to feed each lamb four or five times during the first 18 h after birth.

When first confronted with the notion that lambs require 180 ~ 210 mL colostrum/kg during the first 18 h after birth, some farmers are very sceptical about whether ewes can produce such large volumes of colostrum. The reasons given include the following: the udder of sheep is small (compared with those of goats and dairy cows); such farmers can only extract quite small volumes by hand milking (usually without using a treat lubricant and with the ewe upended between their legs); the volumes of colostrum thought to be necessary to prevent infections are much smaller. However, there is now no doubt that ewes which are healthy and have been fed adequately during late pregnancy can produce the volumes of colostrum indicated above.

The amounts of colostrum available depend mainly on ewe nutriton during late pregnancy and the number of lambs carried. There are also breed differences.

In well-fed ewes (body condition scores of 3 to 4), large volumes of colostrum accumulate in the udder just before birth and large volumes continue to be produced during the first 18 h after birth. Conversely, underfed ewes (condition scores 1.5 ~ 2) have very little colostrum in their udders at birth and it takes several hours for colostrum production to increase. The overall colostrum production by underfed ewes is usually about half that of well-fed ewes during the first 18 h, but in some underfed ewes no colostrum is produced.

When ewes are well fed, more colostrum is produced as the number of lambs carried increases in order to provide enough for the extra lambs. In underfed ewes, however, the presence of twins or triplets is often associated with the production of less colostrum, because a heavy lamb burden usually demands more nutrients than the ewe can supply when feed quality and/or availability are low. In such cases the presence of two or more lambs can cause undernutrition in ewes which would otherwise be eating sufficient, or it can transform what would be moderate undernutrition into severe undernutrition.

Colostrum supply is therefore more likely to be inadequate when the ewe is in poor body condition at lambing, and particularly when she has two or more lambs. These principles apply for all breeds.

3. New Words and Phrases

insufficient [ˌɪnsəˈfɪʃnt] *adj.* 不足的，不够的
beneficial [ˌbenɪˈfɪʃl] *adj.* 有利的，有益的
foster [ˈfɒstər] *v.* 培养；促进；抚育；代养 *adj.* 寄养的；代养的
hypothermia [ˌhaɪpəˈθɜːmiə] *n.* 低体温
immunoglobulin [ɪˈmjuːnəʊˈglɒbjʊlɪn] *n.* 免疫球蛋白
equivalent [ɪˈkwɪvələnt] *adj.* 相等的，相当的，等效的；等价的 *n.* 对等物
diarrhoea [ˌdaɪəˈrɪə] *n.* 腹泻
pathogen [ˈpæθədʒən] *n.* 病菌，病原体
syring [ˈsɑɪrɪŋ] *n.* 注射器，灌注器
sceptical [ˈskeptɪkl] *adj.* 怀疑的；怀疑论者的
lubricant [ˈluːbrɪkənt] *n.* 润滑剂，润滑油 *adj.* 润滑的
upend [ʌpˈend] *v.* 颠倒，倒放
lactose [ˈlæktəʊz] *n.* 乳糖
exocrine [ˈɛksəkrɪn] *adj.* 外分泌的
gut [gʌt] *n.* 内脏
underfed *adj.* 营养不良的
moderate [ˈmɒdərət] *adj.* 温和的；适度的，中等的

4. Notes to the Text A

(1) It is well known that early intakes of colostrum by lambs during the first day after birth improve survival rates and it is obvious that measures which cater for orphans and lambs from ewes with insufficient colostrum would be beneficial.

这是一个复合句，首先是“and”引导两个并列句，然后第二个句子又包括定语从句，“which”引导定语从句修饰“measures”。“It is well known that…”：为“众所周知……”；“it is obvious that…”为“显而易见……”。

(2) Neverthelesss, problems arise because farmers do not know how much colostrum a lamb needs nor the quantities of colostrum which are available for fostered lambs or which are obtainable by milking; nor are they aware that large volumes of colostrum can be obtained from ewes quickly and effectively by hand milking.

这是一个复合句，首先是“because”引导状语从句，状语从句是两个并列句构成，用“;”号分开。

(3) The transition from colostrum to milk is gradual and is accompanied by marked decrease in the concentrations of antibodies and sodium, and by increase in potassium and lactose.

句中“is gradual and is accompanied”并列结果和过程。“by…, and by…”并列表示同时发生。

(4) Lambs fed to appetite by bottle every 2 h can drink about 270 mL colostrum/kg bodyweight (equivalent to 1080 mL for a 4 kg lamb) during the first 18 h after birth.

句中“fed to appetite by bottle every 2 h”分词短语做定语修饰 lambs。

(5) Colostrum availability is affected by breed, ewe nutrition during late pregnancy and the number of lambs born, whereas sucking success depends on the early establishment of a good ewe-lamb bond and on competition between litter mates.

句中“whereas”引导让步状语从句。

(6) In order to avoid hypothermia, lambs born in average field conditions (0~10℃, windy, rainy) need about 210 mL colostrum/kg bodyweight during the first 18 h, while intakes of about 180 mL/kg would be adequate in housed animals (still, dry air at 2~10 ℃).

句中“while”引导让步状语从句。“In order to avoid hypothermia”为目的状语。

(7) These quantities will normally also be sufficient to protect lambs against gut infections which cause watery mouth or diarrhoea because 200 mL of colostrum usually contain enough immunoglobulins for this purpose.

句中“because”引导原因状语从句。

(8) When feeding orphan or sick lambs by stomach tube, it is important to avoid excessive stomach distension, so it is not wise to give more than 50 mL/kg on each occasion.

这是一个复合句，首先是“so”结果状语从句，在主句是“when 加动词 ing”形式，当“when”后面引导的从句和主句主语一致时候，可以省略主语+be 动词。

(9) In underfed ewes, however, the presence of twins or triplets is often associated with the production of less colostrum, because a heavy lamb burden usually demands more nutrients than the ewe can supply when feed quality and/or availability are low.

句中“because”引导原因状语从句，“however”插入语。

5. Exercises

(1) Fill in the blanks by finishing the sentences according to the passage.

①Colostrum usually ________ in the udder during the final few days of pregnancy.

②The usual definition of ________ emphasizes a high immunoglabium content, but in this chapter the word ________ will be used simply to mean early milk, i. e. all the exocrine secretions produced by the udder during the first 24 h after birth.

③Lambs fed to appetite by bottle every 2 h can drink about ________ mL colostrum/kg bodyweight (equivalent to 1080 mL for a 4 kg lamb) during the first 18 h after birth.

④During bad weather (cold, wind and, or rain) the lamb must produce more heat to avoid ________, its requirement for colostrum increases.

⑤These quantities will normally also be sufficient to protect lambs against gut infections which cause watery mouth or diarrhoea because 200 mL of colostrum usually contain enough ________ for this purpose.

⑥These colostrum requirements represent the total ________ during the first 18 h after birth.

⑦The amounts of colostrum available depend mainly on ewe nutriton during late ________ and the number of lambs carried.

(2) Answer the following questions according to the passage.

①What is colostrum?

②Why need lambs during the first day after birth intake colostrum?

③When does colostrum accumulate in the udder during pregnancy?

④What does the amount of colostrum a lamb drinks depends on?

⑤How much colostrum do lambs need about during the first 18 h in average field conditions (0~10 ℃, windy, rainy)?

⑥Why is it important to avoid excessive stomach distension when feeding orphan or sick lambs by stomach tube?

⑦How can large volumes of colostrum be accumulated in the udder just before birth?

⑧When is colostrum supply more likely to be inadequate?

(3) Translate the following sentences into English.

①Neverthelesss, problems arise because farmers often do not know how much colostrum a lamb needs nor the quantities of colostrum which are available for fostered lambs or which are obtainable by milking; nor are they aware that large volumes of colostrum can be obtained from ewes quickly and effectively by hand milking.

②The transition from colostrum to milk is gradual and is accompanied by marked decrease in the concentrations of antibodies and sodium, and by increase in potassium and lactose.

③ Colostrum availability is affected by breed, ewe nutrition during late pregnancy and the number of lambs born, whereas sucking success depends on the early establishment of a good ewe－lamb bond and on competition between litter mates.

④These figures must be multiplied by the lamb's weight (kg) in order to estimate the total volume of colostrum required during the first 18 h after birth.

⑤These quantities will normally also be sufficient to protect lambs against gut infections which cause watery mouth or diarrhoea because 200 mL of colostrum usually contain enough immunoglobulins for this purpose.

⑥However, there is now no doubt that ewes which are healthy and have been fed adequately during late pregnancy can produce the volumes of colostrum indicated above.

⑦When ewes are well fed, more colostrum is produced as the number of lambs carried increases in order to provide enough for the extra lambs.

课文A 初乳

众所周知，羔羊在出生后第一天就哺喂初乳可以提高存活率，显而易见，满足孤羔和母羊初乳不足的羔羊需要的措施是非常有益的，然而，在养殖过程会遇到一些问题，例如养殖户们往往不清楚一只羔羊需要多少初乳，也不知道培育羔羊提供的初乳量和挤乳获得初乳的量，他们也没有意识到手工挤乳能够快速而有效地从母羊身上挤出大量的初乳。

初乳通常积累在怀孕最后几天的乳房中，在出生后12~24h内，初乳开始分泌，但是随着产乳量的增加，初乳逐渐被稀释且转变为常乳，抗体和钠的浓度显著降低，钾和乳糖的浓度增加。一般对于初乳的定义着重强调初乳中大量的免疫球蛋白，但是在本单元中，“初乳”仅仅简单地定义为早期乳。即在羔羊出生后最初的24h内乳房所分泌的乳汁。

刚出生18h的羔羊每隔2h通过奶瓶哺乳，满足食欲，大约270mL/kg

(相当于一个 4kg 的羔羊能喝 1080mL 的初乳)，这是羔羊的最大摄入量。

一头羔羊所饮的初乳量取决于可获得的乳量和能否成功的吮吸，可获得的乳量受羊的种类、怀孕末期母羊的营养水平和产仔数的影响。然而能否成功地吮吸取决于产后尽早地建立一个良好的亲子关系，还取决于幼仔们之间的相互竞争。显而易见，不同羔羊对初乳的吸收量在零和最大摄入量之间变动。一头羔羊所需要的初乳量主要取决于它需要多少热量。因此，能增加产热量的任何因素都会增加初乳的需要量。在恶劣天气下（寒冷、风、雨）羔羊必须产生更多的热量来避免体温降低，所以就需要更多的初乳。

为了避免体温下降，在室外条件下（0~10℃，有风、下雨）出生后 18h 的羔羊需要 210mL/kg 初乳，而在室内条件下（在 2~10℃，无风，干燥），出生后 18h 的羔羊需要 180ml/kg 初乳就能满足需要。为了估计出生后 18h 羔羊所需的总初乳量，数据必须乘体重。

在正常情况下，由于初乳中含有免疫球蛋白，可以完全保护羔羊，避免肠道感染或者腹泻。初乳中的免疫球蛋白对特定的病原体具有免疫特性，如果没有免疫特异性，羔羊将得不到保护。当羔羊被喂养了不含免疫球蛋白的牛乳作为替代品，那么至少在羔羊出生后的第一天要给它们提供 200mL 的初乳来减少早期生长过程中肠道感染和其他感染的风险。

这些初乳的吸收量表明羔羊在出生 18h 后的总摄入量。当通过导流管哺喂孤羔和病羔的时候，避免过量胃胀非常重要。所以每次哺喂超过 50mL/kg 是非常不可取的。用 50mL 的注射器给羔羊吃乳，这样既节省时间又简单可行，这个操作就是一头羔羊出生后 18h 羔羊每次哺喂按每体重（kg）计算用 50mL 注射器需要 4 或 5 次是非常有必要的。

当养殖户们面对着每头羔羊在出生后 18h 要满足 180~210mL/kg 的初乳这个问题，他们很多人怀疑母羊是否能产那么多量的初乳。他们怀疑的理由有以下几点：绵羊的乳房很小（与山羊、奶牛的乳房相比）；那么养殖户们只能用手额外挤出一点乳（通常不用乳头润滑剂，并且把母羊倒放置在他们腿之间）。初乳量只需很少就能防止感染。但是毫无疑问健康的母羊或者是在怀孕后期饲喂营养水平较高的母羊能产生更多的乳。

可产生的初乳量主要取决于母羊怀孕后期的营养水平以及所怀的胎数，其次还受品种类型不同的影响。

饲喂好的母羊（身体评分在 3~4 分），在产前就有大量的初乳储存在乳房中，然后在产后 18h 内初乳继续分泌。相反，如果饲喂状况不好的母羊（身体评分在 1.5~2 分）在分娩后几乎没有初乳，初乳量的增加要花费几个小时，弱母羊产生的总初乳量通常在产后18h之内是健康母羊的一半，甚至不产初乳。

母羊营养条件好，就能产生更多初乳，它可以为别的羔羊提供足够的初乳，哺乳的羔羊数增多。可是营养不良的母羊，产完双胞胎或三胞胎的羔羊后，产生的初乳更少。因为当营养不良或者饲料利用率低的时候，产仔数越多，羔羊所需营养远远大于母羊所提供的初乳，在那样的条件下，产两头或更多的羔羊可以造成一直营养充足的母羊营养不良，也可以使中度营养不良变成严重营养不良。

由于母羊产后身体虚弱，特别使在分娩多仔羔羊的情况下，初乳的供应可能就更加不充足了，这准则适用于山羊和绵羊所有种类。

Part Two Extensive Reading

1. Text B Problems with Newborn Lambs

In addition to making sure the lambs are getting adequate feed, there are some other problems to watch for. Hypothermia is one of the most common problems facing newborns, and it doesn't have to be all that cold for hypothermia to occur. In fact, a lamb can suffer from starvation hypothermia on a moderately warm, sunny day.

Another fairly common problem is a "weak" lamb, which may result if the ewe had a long, difficult delivery.

Hypothermia

Guard newborns and young lambs against hypothermia, which is implicated in about half of all lamb deaths. Hypothermia has two basic causes: exposure and starvation. As it implies, exposure hypothermia is a result primarily of extremely cold temperatures or cold temperatures mixed with drafts. This can kill wet lambs within the first few hours of birth. Starvation hypothermia can occur in lambs from 4 and 5 hours old to a couple of days old.

Once dry and fed, lambs can withstand quite low temperatures, but due to a large ratio of skin area to body weight, wet or hungry lambs can chill quite quickly. A hypothermic lamb will appear stiff and be unable to rise. Its tongue and mouth will feel cold to the touch. You must warm it immediately with an outside heat source, because it has lost its ability to control its temperature. Wrapping it in a towel or blanket will not suffice.

There are several methods of warming lambs. Water warming is probably the best choice for very cold lambs, with air warming a close runner-up. Some people use infrared lamps, but this is probably the least desirable method because the infrared lamps can seriously burn the lamb and can cause fires if used in the lambing shed. If you plan to lamb in winter, consider buying or building a lamb-warming box. But if you'll normally lamb in the spring, then you can probably get away with bringing the occasional cold lamb into the house. Warm it in a big cardboard box or do it like we did, and pick up an old playpen (yeah, you remember those prisons we were subjected to as toddlers) at a flea market. Your box or playpen can be set near a woodstove or a heat vent or in front of the oven (with the door open) to warm the air and the lamb. A blow dryer also helps, but again, be careful not to burn the lamb.

If you are dealing with a slightly older lamb that has become severely hypothermic from starvation, then it will need an injection of glucose. This happens to lambs that don't get fed within an hour or so after birth. They need the energy in that first hit of milk very badly. If they don't get it, their body temperature begins to fall, and they begin to "feed" off the glycogen, or sugar, reserves in their body. Without the glucose injection, a lamb will die during warming.

The use of a plastic "lamb coat" in cold weather can be beneficial because it helps the lamb retain a great deal of body heat. A newborn lamb appears wrinkly because there is very little body fat under the skin. It takes 3 to 5 days to build up that fat layer under the skin, which acts as natural insulation. When a lamb coat is used to help the lamb retain body heat, the energy that would be used to keep it warm is converted to body fat. This can be especially beneficial to twins and triplets with marginal milk intake.

Weak Lambs

A lamb that has been weakened by a protracted or difficult birth may be suffering from anoxia (lack of oxygen) or have fluid in its lungs. The first few minutes are critical. If it gurgles with the first breaths or has trouble breathing, swing it as discussed previously. Two or three swings normally get things going. Be sure that you have a firm grasp on the lamb (the lamb will be slick) and that there are no obstructions in the path of your swing.

It is not essential for the first feeding to be colostrum, but make sure the lamb does receive colostrum during the first few hours of life. The lamb's ability to absorb the antibodies in the colostrum drops rapidly from birth to approximately 16 hours of life. For a very weak lamb, you may have to give the first feeding from a baby bottle with the nipple hole enlarged to about the size of a pinhead or use a stomach tube to feed. Give 59 mL of warmed colostrum to give the lamb strength. Do not force the lamb, if it has no sucking impulse, the milk will go into its lungs and cause death. Often, a weak lamb can get up on its feet after just one bottle feeding (or stomach-tube feeding) and be ready to nurse from its mother without further assistance.

2. Notes to the Text B

If they don't get it, their body temperature begins to fall, and they begin to "feed" off the glycogen, or sugar, reserves in their body.

复合句，"if" 引导条件状语从句，"and" 引导并列句。

3. Answer Question

(1) What is one of the most common problems facing newborns?

(2) How to guard newborns and young lambs against hypothermia?

(3) What may a weak lamb be suffering from?

课文B　新生羔羊的问题

除了要保证羔羊能吃到足够的饲料，还有一些问题需要注意，低温是新生羔羊最常见的问题，并不是所有的低温都由寒冷造成，事实上，一只羔羊也可能在温暖的有阳光的日子里忍受饥饿造成的低温。

另外一个相当常见的问题就是由生产时间长而造成难产所生下的弱羔。

体温过低

保护新生羔羊和幼年羔羊免受体温过低的伤害，体温过低所造成的死亡率高达50%。造成体温过低的两个基本原因：暴露和饥饿。暴露低温主要是寒冷的温度或者是有寒风的低温造成的，这会杀死刚出生几个小时的全身湿漉漉的羔羊，饥饿低温主要发生在出生后4~5h到2d之中。

一旦处在干燥、饲喂充足的条件下，羔羊能忍受低温，但是由于皮肤表面积占体重比例大，潮湿或者饥饿，羔羊可能很快打寒战，低温使羔羊身体僵直，不能站立，它的舌头和嘴将会摸起来很凉，必须用一个外部的热源立刻使它暖和起来，因为它已经失去了调控体温的能力，用毛巾或毯子裹住它就可以了。

有很多方法都可以让羔羊暖和起来，用水温暖寒冷的羔羊是非常好的选择，用空气温暖则是第二选择。一些人用红外线灯给羔羊加热，但是这些可能是最不理想的方法，因为它们可能灼伤羔羊，如果被用在羊棚里可能造成火灾。如果计划冬天产羔羊，可以考虑买或建造一个羔羊的暖箱，如果是按正常时间春天产羔，也可以把偶尔感到寒冷的羔羊带入这间房子，在大纸盒里保暖羔羊，或者按我们以前看护婴幼儿那样，从跳蚤市场买一个旧围栏，就是那种保护学步的幼儿用的围栏。盒子或者护栏要设置在热通气孔附近，或者炉子前面，但炉子门是开着的，可以加热空气帮助羔羊保暖。吹风机也挺有用，但再一次警告大家，不要烧着羔羊。

如果你正在解决稍大的羔羊因为饥饿而造成低温的话，给它注射一些葡萄糖，羔羊发生这种情况主要是因为出生后一个小时都没有喂养。它们非常需要初乳给它们提供能量，如果它们没有吃到初乳，它们会消耗自身储存的糖原等物质，体温开始下降。没有注射葡萄糖，羔羊会在给它保暖的期间死去。

在寒冷的天气里，可以给羔羊穿上塑料外套来帮助羔羊保存身体热量。新生羔羊身上有很多皱褶，是因为皮下几乎没有脂肪。而在3~5d后皮下脂肪变得发达，免受自然界因素的影响。穿上一件外套可以帮助羔羊保持体温，本应该用来维持体温的能量转变为脂肪，这对只吸收了刚满足需要的乳量的双羔或三羔是非常有利的。

弱羔

产程长或经历难产的羔羊容易缺氧，刚出生的几分钟是最关键的，如果在出生后开始咩叫时能听到咕噜声，或者有呼吸障碍，要按前述那样抓羔羊，正常情况下摇动2~3次就可以使它们呼吸顺畅，一定要保证抓紧羔羊，摇动时不要有障碍物阻挡。

第一次吃乳没有必要是初乳，但是要确保在出生后的头几个小时能够喝到初乳。羔羊吸收初乳抗体的能力从出生到出生后大约16h快速地下降，对一个体质非常弱的羔羊来说，你可以把奶嘴扎大一些达到平头针大小的孔，来给它第一次喂乳，或者用胃导管给它第一次喂乳，喂59mL的温暖初乳使它有力量，如果羔羊没有吸吮的冲动，不要强制羔羊吸，否则的话会进入肺部造成死亡，很常见的就是在给一头弱羔，喂了一瓶乳之后，这头羔羊自己就能站立，或者是不需要任何帮助开始自行吃母乳。

扫码进行拓展学习

Lesson 20
Rabbit

Part One Intensive Reading

1. Learning Objective

After learning this lesson, you should understand the following:

(1) Describe how to proper handling of your rabbit.

(2) Describe why rabbit are prey animals.

(3) Give an overview of caring for rabbit.

(4) Explain why rabbit are prone to multiple dental problems.

(5) Give an overview of how to keep the rabbit comfortable and healthy.

2. Text A Management of Rabbit

Handling

Proper handling of your rabbit is essential. Always support the hind end of your rabbit when you pick it up. Never pick up a rabbit by its legs or ears. Rabbits can seriously injure their backs when picked up without supporting their hind end! Handle your rabbit often when they are young to increase their acceptance of affection when they are older. I do not recommend letting children under the age of 12 handle rabbit without supervision. The best way for young children to handle rabbit is to sit on the floor and let the rabbit lay in the lap of the child with legs crossed.

Rabbits are very curious critters that love to explore by chewing on objects in their environment. Unfortunately many rabbits have been injured by chewing on electrical wires or other dangerous items. Always keep your rabbit caged when you are not there to supervise her activity. Wire cages are readily available and work well by allowing the feces and urine to fall through the cage. However, always provide a solid surface for your rabbit to sit or lie on if she chooses. A rabbit forced to sit on wire can develop "sore hocks" which is a skin infection that can become very serious if left untreated. Keep the cage and the bedding as clean as possible. Rabbits that are allowed to come into contact with their urine for long periods can also develop "sore hocks". This problem is easier to prevent than it is to treat (not to mention very uncomfortable for your rabbit)!

Line your rabbit's cage with newspaper, shredded paper, or a paper based bedding/litter available at a commercial pet store. Eco-bedding brand (which looks like crinkled brown paper) is an excellent choice. DO NOT USE WOOD SHAVINGS of any kind.

Rabbit can be litter box trained! Begin with the rabbit and the box in a small area. Place the box in a location your rabbit normally defecates. Once she learns to use the box regularly, you can gradually increase the area she is allowed to explore. If she has access to more than one room, place a box in each room. Males and females can be trained to use a litter box. Rabbit litter boxes are not the same as cat litter boxes. Do not use cat litter for rabbit.

Diet

Fresh water should always be available. A sipper bottle or water bowl raised off the cage floor is more sanitary than a bowl that sits on the floor. In the past, rabbits were fed unlimited amounts of a pellet-based diet. Research and experience indicate that this is not necessary and can even be harmful. Pelleted diets have been implicated in soft, pasty stools, hairballs, obesity and its related diseases, and urinary diseases. In addition, rabbit that eat only pellets and do not have fresh hay to chew can develop painful overgrown teeth. In the wild, rabbit have adapted to eating large amounts of food that make up relatively poor diets, *e. g.* grasses, tree bark, and other vegetation. Pellets are too rich for them. Limit your rabbit's intake of pellets to only 1/8 cup of a timothy hay-based product per 2.3 kg of body weight per day (About 30 g). Please do not feed the rabbit pellet mixes with pretty seeds and colored pellets. This is like candy to them and can cause stomach issues. Please feed only timothy pellets, like those offered by Oxbow. An unlimited amount of fresh, clean and dry (non-moldy) timothy hay should always be available. Alfalfa hay is not recommended if over 4 months of age. It is too high in calcium and fat for teenage to adult rabbit.

Provide fresh greens on a daily basis. Your rabbit should receive one heaping cup per 2.3 kg of body weight of a mixture of at least three greens: red leaf lettuce, green leaf lettuce, dandelion greens, carrot tops, parsley, romaine lettuce, or fresh picked non chemically treated grass. Be sure to vary the greens you give. Do not give the same ones each week. Providing a variety will ensure a more balanced diet. Spinach and kale can also be given on occasion but sparingly.

Spaying and Neutering

Spaying (removal of ovaries and uterus) and neutering (removal of testicles) are very important for your pet rabbit. Spaying your female prevents uterine cancer and neutering your male prevents testicular cancer. Both of these cancers are frequently seen in middle-aged to older intact rabbits. Spayed or neutered rabbits also tend to be better pets due to fewer aggressive and sexual behaviors.

Teeth

Unfortunately, rabbits are prone to multiple dental problems. One important way to avoid this is to provide plenty of fresh timothy or orchard grass hay for your rabbit to chew daily. Eating hay helps to wear down the teeth naturally, avoiding painful overgrowth of the molars. A rabbit that is not eating is a serious problem. In addition, you should regularly check your rabbits' front teeth for overgrowth. Some rabbits have front teeth that do not align properly, causing them to grow abnormally long.

Nail Trims

When clipping nails, it is helpful to wrap your rabbit in a towel (a bunny burrito). The towel will have to be wrapped snuggly. Each foot can be individually pulled out of the towel for clipping. Nail clipping can be done with either dog or human nail clippers.

3. New Words and Phrases

curious [ˈkjʊəriəs] *adj.* 好奇的；奇妙的；好求知的；稀奇的

critter [ˈkrɪtər] *n.* 动物

chew [tʃuː] *vt. & vi.* 咀嚼，咬；深思，考虑 *n.* 咀嚼；咀嚼物

shred [ʃrɛd] *n.* 碎片；破布；少量 *vt. & vi.* 撕碎，切碎；用碎纸机撕毁（文件）

defecate [ˈdefəkeɪt] *v.* 澄清 *vt.* 排便

sanitary [ˈsænətri] *adj.* 卫生的；清洁的 *n.* 公共厕所

implicate [ˈɪmplɪkeɪt] *vt.* 暗示；牵涉，涉及（某人）；表明（或意指）……是起因 *n.* 包含的东西

hairball [ˈherˌbɔːl] *n.*（动物在胃或肠积成的）毛团

urinary [ˈjʊrəneri] *adj.* 尿的；泌尿器的；泌尿的；*n.* 小便池

vegetation [ˌvɛdʒɪˈteʃən] *n.* 植物（总称），草木；赘生物，增殖体

dandelion [ˈdændlˌaɪən] *n.* [植] 蒲公英

parsley [ˈpɑrsli] *n.* 西芹，欧芹；洋芫荽

spinach [ˈspɪnɪtʃ] *n.* 菠菜

spay [speɪ] *vt.* 切除卵巢

neuter [ˈnuːtər] *adj.* 中性的，不及物的，（生物）无性的 *n.* 中性名词，无性动物，阉割动物

uterine [ˈjutərɪn] *adj.* 子宫的，同母异父的，母系的

testicle [ˈtɛstɪkəl] *n.* 精巢；睾丸

4. Notes to the Text A

（1） Always keep your rabbit caged when you are not there to supervise her activity.

句中“caged”在这里翻译为“关入笼子中”。“when”引导时间状语从句。

（2） A rabbit forced to sit on wire can develop “sore hocks” which is a skin infection that can become very serious if left untreated.

句中“which”引导定语从句。“that”引导定语从句，在从句中做主语。

（3） A sipper bottle or water bowl raised off the cage floor is more sanitary than a bowl that sits on the floor.

句中“raised”在这里修饰“water bowl”，意思是“提起的水碗”。

5. Exercises

（1） Fill in the blanks by finishing the sentences according to the passage

①Handle your rabbit often when it is young to increase its acceptance of ________ when it is older.

②Rabbits are very curious critters that love to explore by ________ on objects in their environment.

③Once she learns to use the box regularly, you can gradually increase the area she is allowed to ________ .

④Spinach and kale can also be given on ________ but sparingly.

⑤ Eating hay helps to wear down the teeth naturally, avoiding Painful ________ of the molars.

（2） Answer the following questions according to the passage

①What's the best way for young children to handle rabbit?

②How to line with the rabbits' cage?

③Why do not feed the rabbit pellet mixes with pretty seeds and colored pellets?

④What are very important for your pet rabbit health?

⑤What is the important way to avoid multiple dental problems of rabbit?

(3) Translation of the following sentences into Chinese

①Wire cages are readily available and work well by allowing the feces and urine to fall through the cage. However, always provide a solid surface for your rabbit to sit or lie on if she chooses.

②In the past, rabbits were fed unlimited amounts of a pellet - based diet. Research and experience indicate that this is not necessary and can even be harmful.

③In addition, rabbit that eat only pellets and do not have fresh hay to chew can develop painful overgrown teeth.

④An unlimited amount of fresh, clean and dry (non - moldy) timothy hay should always be available.

⑤Some rabbits have front teeth that do not align properly, causing them to grow abnormally long.

课文A 家兔管理

处理

掌握正确抓握兔子的方法是很有必要的。当你抓起兔子的时候，一定要支撑它的后端。千万不要抓兔子的腿或耳朵。兔子被抓起来的时候，如果其后腿没有支撑，会严重伤害它们的背部！在兔子幼小的时候经常抚摸，以增加它长大后的接受度。我不建议让12岁以下的孩子，在没有监护的情况下玩兔子。对于小孩子来说，最好的办法就是坐在地板上双腿交叉，让兔子躺在孩子的膝盖上。

兔子是非常好奇的动物，喜欢通过咀嚼周围的东西来探索。不幸的是，许多兔子因为咬电线或其他危险物品而受伤。当你不在附近，无法管理你的兔子活动时，一定要把它关在笼子里。铁丝笼容易得而好用，粪便和尿液通过笼底排出。然而，经常提供一个坚实的表面，可以为你的小兔子选择坐或躺。兔子被迫坐在钢丝上会出现“关节溃疡”，这是一种皮肤感染，如果不及时治疗，可能会变得非常严重。尽可能保持笼子和床上用品的清洁。兔子长时间接触其尿液也会发生“关节溃疡”。这个问题预防比治疗（更不用说让你的兔子非常不舒服）更容易！

在兔子的笼子里，用报纸、碎纸片或商业宠物店出售的纸质床上用品为小兔做窝。生态床上用品品牌（看起来像皱巴巴的牛皮纸）是一种很好的选择。不要用任何木屑。

可以训练兔子使用砂盆！把兔子和砂盆放在小范围内开始。把砂盆放在兔子通常排便的地方。一旦它学会经常使用砂盆，你可以逐渐增加它的活动区域。如果它能进入多个房间，就在每个房间里放一个砂盆。可以训练雄性和雌性使用同箱。兔子砂盆和猫砂盆是不一样的。不要用猫砂盆给兔子。

日粮

保证随时可饮用到新鲜的水。水壶或水碗离开笼子地板的比放在地板上卫生。过去，喂兔方法是自由采食颗粒饲料。研究和经验表明，这是不必要的，甚至可能是有害的。这种方法与软粪便、拉稀、（胃中聚成）毛团、肥胖及其相关的疾病有关，也和泌尿系统疾病有关。此外，兔子只吃颗粒饲料，没有新鲜的干草可以咀嚼，易使牙齿过度生长、痛苦不堪。在野外，兔子已经适应了大量的采食，野外食物由营养相对贫乏的原料构成，例如草、树皮和其他植物。颗粒料对它们来说营养太丰富了。限制你的兔子每天的颗粒摄入量，以每 2.3kg 体重只能摄入 1/8 杯梯牧草干草为基础的产品（约 30g）。请不要给兔子喂混有漂亮种子和彩色颗粒的饲料，这对它们来说就像糖果，会引起胃病，请只给喂食梯牧草，就像 Oxbow 提供的那种。不限量的新鲜、干净和干燥（非发霉）的梯牧草可经常使用。如果超过 4 个月大，不推荐使用苜蓿干草，对于仔幼兔和成年兔子来说，它的钙和脂肪含量太高了。

每天提供新鲜的绿草。你的兔子每 2.3kg 体重应该得到一杯堆积的草，里面至少有三种绿色蔬菜的混合物：红叶莴苣、绿叶莴苣、蒲公英叶、胡萝卜叶、欧芹、生菜或者新摘的未经化学处理的草。一定要保证你的鲜草多样化。每周不要给同一种草。提供多样化的饮食将确保日粮平衡。菠菜和甘蓝也可以偶尔饲喂，但要有节制。

阉割和绝育

对于宠物兔子来说，绝育（卵巢和子宫的切除）和阉割（睾丸的切除）是非常重要的。为雌性做绝育手术可以预防子宫癌，为雄性做绝育手术可以预防睾丸癌。这两种癌症常见于未切除的中老年兔子。由于攻击性和性行为缺失，阉割或绝育的兔子往往是更好的宠物。

牙齿

不幸的是，兔子有多种牙齿问题。避免这种情况的一个重要方法是给你的兔子足够的新鲜梯牧草或果园草干草，保证每天咀嚼。吃干草可以自然磨损牙齿，避免臼齿过度生长。没有充分咀嚼对兔子是个严重的问题。另外，你应该定期检查你的兔子的门牙是否过度生长。有些兔子的门牙排列不整齐，导致它们长得不正常。

指甲修整

修整指甲的时候，用毛巾（把小兔子卷起来）把兔子包起来是很有帮助的。毛巾必须紧紧地裹起来。每只脚都可以单独地从毛巾里抽出来用来剪指甲，用狗或者人的指甲刀都可以。

Part Two Extensive Reading

1. Text B Caring of Rabbit with Special Need

Caring for special needs rabbits can be a sensitive issue. Some people feel that keeping disabled rabbit alive is cruel and inhumane. Those who devote themselves to a special needs rabbit and are rewarded by seeing it exhibit the behavior of a happy or content rabbit feel beyond doubt that they have made the correct choice. I know I would not trade the times my special needs rabbit hums around the room and even tries to kick his heels for anything.

Special needs rabbit include those suffering from a variety of diseases and conditions. Splayleg, stroke, broken back, maloccluded teeth, chronic bacterial infection, recurring abscesses, cancer, and encephalitozoonosis are just a few of the conditions that may require a rabbit to have long-term special care. The specific needs of the rabbits will vary depending on the nature of the condition.

2. Notes to the Text B

Those who devote themselves to a special needs rabbit and are rewarded by seeing it exhibit the behavior of a happy or content rabbit feel beyond doubt that they have made the correct choice.

句中“Who”在这里引导主语从句；“that”引导宾语从句。

3. Answer Questions

(1) Why can caring for special needs rabbit be a sensitive issue?

(2) What do special needs rabbit include?

（3）What will the specific needs of the rabbit vary depending on?

课文 B　照顾有特殊需要兔子的建议

照顾有特殊需要的兔子是个敏感问题。有些人觉得让残疾的兔子活着是残忍和不人道的。但那些把自己奉献给有特殊需要的兔子，并因看到它表现出快乐或满足的行为而感觉得到奖励的人，毫无疑问地感到他们做出了正确的选择。我知道我不会让兔子在房间里嘶叫，或者试图踢兔子的脚后跟来满足我的特殊需要。

有特殊需要的兔子包括那些患有各种疾病的兔子。外翻脚、中风、背部骨折、牙齿咬合不良、慢性细菌感染、复发性脓肿、癌症，以及颅内寄生虫病，这些只是需要长期特殊照顾的几种情况。兔子的特殊需求取决于身体的状况。

扫码进行拓展学习

Lesson 21

Animal Body

Part One Intensive Reading

1. Learning Objective

After learning this lesson, you should understand the following:

(1) Explain where epithelial tissue covers and it's functions.

(2) Explain what connective tissue serves in the body.

(3) Describe muscle tissue facilitates movement of the animal by contraction of individual muscle cells.

(4) Give an overview ofthe neuron is the functional unit of the nervous system.

2. Text A Organization of the Animal Body

Organs in animals are composed of a number of different tissue types.

Epithelial Tissue

Epithelial tissue covers body surfaces and lines body cavities. Functions include lining, protecting, and forming glands. Three types of epithelium occur: ①Squamous epithelium is flattened cells. ②Cuboidal epithelium is cube-shaped cells. ③Columnar epithelium consists of elongated cells.

Any epithelium can be simple or stratified. Simple epithelium has only a single cell layer. Stratified epithelium has more than one layer of cells. Pseudostratified epithelium is a single layer of cells so shaped that they appear at first glance to form two layers.

Functions of epithelial cells include: ① Movement of materials in, out, or around the body. ② Protection of the internal environment against the external environment. ③ Secretion of a product.

Glands can be single epithelial cells, such as the goblet cells that line the intestine. Multicellular glands include the endocrine glands. Many animals have their skin composed of epithelium. Vertebrates have keratin in their skin cells to reduce water loss. Many other animals secrete mucus or other materials from their skin, such as earthworms do.

Connective Tissue

Connective tissue serves many purposes in the body, including: ① binding. ②supporting. ③protecting. ④forming blood. ⑤storing fats. ⑥filling space.

Connective cells are separated from one another by a non-cellular matrix. The matrix mav be solid (as in bone), soft (as in loose connective tissue), or liquid (as in blood). Two types of connective tissue are Loose Connective Tissue (LCT) and Fibrous Connective Tissue (FCT). Fibroblasts are separated by a collagen fiber-containing matrix. Collagen fibers provide elasticity and flexibility. LCT occurs beneath epithelium in skin and many internal organs, such as lungs, arteries and the urinary bladder. This tissue type also forms a protective layer over muscle, nerves, and blood vessels.

Adipose tissue has enlarged fibroblasts storing fats and reduced intracellular matrix. Adipose tissue facilitates energy storage and insulation.

Fibrous Connective Tissue has many fibers of collagen closely packed together. FCT occurs in tendons, which connect muscle to bone. Ligaments are also composed of FCT and connect bone to bone at a joint.

Cartilage and bone are "rigid" connective tissues. Cartilage has structural proteins deposited in the matrix between cells. Cartilage is the softer of the two. Cartilage forms the embryonic skeleton of vertebrates and the adult skeleton of sharks and rays. It also occurs in the human body in the ears, tip of the nose, and at joints.

Bone has calcium salts in the matrix, giving it greater strength. Bone also serves as a reservoir (or sink) for calcium. Protein fibers provide flexibility while minerals provide elasticity. Two types of bone occur. Dense bone has osteocytes (bone cells) located in lacunae connected by canaliculi. Lacunae are commonly referred to as Haversian canals. Spongy bone occurs at the ends of bones and has bony bars and plates separated by irregular spaces. The solid portions of spongy bone pick up stress.

Blood is a connective tissue of cells separated by a liquid (plasma) matrix. Two types of cells occur. Red blood cells (erythrocytes) carry oxygen. White blood cells (leukocytes) function in the immune system. Plasma transports dissolved glucose, wastes, carbon dioxide and hormones, as well as regulating the water balance for the blood cells. P1atelets are cell fragments that function in blood clotting.

Muscle Tissue

Muscle tissue facilitates movement of the animal by contraction of individual muscle cells (referred to as muscle fibers). Three types of muscle fibers occur in animals (the only taxonomic kingdom to have muscle cells): ①skeletal (striated). ②smooth. ③cardiac.

Muscle fibers are multinucleated, with the nuclei located just under the plasma membrane. Most of the cells are occupied by striated, thread-like myofibrils. Within each myofibril there are dense Z lines. A sarcomere (or muscle functional unit) extends from Z line to Z line. Each sarcomere has thick and thin filaments. The thick filaments are made of myosin and occupy the center of each sarcomere. Thin filaments are made of actin and anchor to the Z line.

Skeletal (striated) muscle fibers have alternating bands perpendicular to the long axis of the cells. These cells function in conjunction with the skeletal system for voluntary muscle movements. The bands are areas of actin and myosin deposition in the cells.

Smooth muscle fibers lack the banding, although actin and myosin still occur. These cells function in involuntary movements and/or autonomic responses (such as breathing, secretion, ejaculation, birth, and certain reflexes). Smooth muscle fibers are spindle shaped cells that form masses. These fibers are components of structures in the digestive system, reproductive tract, and blood vessels.

Cardiac muscle is a type of striated muscle found only in the heart. The cell has a bifurcated (or forked) shape, usually with the nucleus near the center of the cell. The cells are usually connected to each other by intercalated disks.

Nervous Tissue

Nervous tissue functions in the integration of stimulus and control of response to that stimulus. Nerve cells are called neurons. Each neuron has a cell body, an axon, and many dendrites. Nervous tissue is composed of two main cell types: neurons and glial cells. Neurons transmit nerve messages. Glial cells are in direct contact with neurons and often surround them.

The neuron is the functional unit of the nervous system. Humans have about 100 billion neurons in their brain alone! While variable in size and shape, all neurons have three parts. Dendrites receive information from another cell and transmit the message to the cell body. The cell body contains the nucleus, mitochondria and other organelles typical of eukaryotic cells. The axon conducts messages away from the cell body.

3. New Words and Phrases

cavity [ˈkævɪtɪ] *n.* 腔；窝洞；蛀牙，龋洞
epithelium [ˌepɪˈθiːlɪəm] *n.* 上皮，上皮细胞
stratiffed [ˈstrætifaid] *adj.* 成层了的，分层的
intestine [inˈtestin] *n.* 肠
keratin [ˈkerətin] *n.* 角蛋白
secrete [siˈkriːt] *vt.* 分泌
mucus [ˈmjuːkəs] *n.* 黏液，胶
earthworm [ˈɜːθwɜːm] *n.* 蚯蚓
matrix [ˈmeitriks] *n.* 基质
fibroblast [ˈfaɪbrəʊblæst] *n.* 纤维原细胞，成纤维细胞
collagen [ˈkɒlədʒən] *n.* 胶原，胶原质
elasticity [ilæˈstisiti] *n.* 弹性，弹力，灵活性
flexibility [fleksɪˈ biliti] *n.* 柔性，灵活性
artery [ɑːrtəri] *n.* 动脉，干道，主流
intracellular [ˌɪntrəˈseljʊlə] *adj.* 细胞内的
facilitate [fəˈsɪlɪteɪt] *vt.* 帮助，使容易，促进
insulation [ɪnsjʊˈleɪʃən] *n.* 绝缘，隔离，孤立，绝缘或隔热的材料，隔声
tendon [ˈtendən] *n.* 腱，筋
ligament [ˈlɪgəmənt] *n.* 韧带
cartilage [ˈkɑːrtɪlɪdʒ] *n.* 软骨
osteocyte [ˈɒstɪəʊˌsaɪt] *n.* 骨细胞
canaliculi [ˌkænəˈlikjuliː] *n.* 小管，细管，微管 canaliculus 的复数
erythrocyte [ɪˈrɪθrəsaɪt] *n.* 红细胞

leukocyte [ˈljuːkəʊsaɪt] *n.* 白细胞
platelet [ˈpleitlit] *n.* 血小板
multinucleate [ˌmʌtiˈnjuːkliit] *adj.* 多核的
myofibril [ˌmaɪəˈfaɪbrəl] *n.* 肌原纤维
sarcomere [ˈsɑːkəʊmɪə] *n.* 肌原纤维节，肌小节
myosin [ˈmaɪəʊsɪn] *n.* 肌浆球蛋白，阻凝蛋白
actin [ˈæktɪn] *n.* 肌动蛋白，肌纤蛋白
cardiac [ˈkɑːdɪæk] *adj.* 心脏的
ejaculation [ɪˌdʒækjʊˈleɪʃən] *n.* 射精
neuron [ˈnjʊərɒn] *n.* 神经细胞，神经元
axon [ˈæksɒn] *n.* 轴突
dendrite [ˈdendraɪt] *n.* 枝状突起；树突
epithelial tissue 上皮组织
squamous epithelium 扁平上皮
cuboidal epithelium 立方上皮
columnar epithelium 柱形上皮
at first glance 乍一看，看一眼
goblet cells 杯状细胞
endocrine gland 内分泌腺
connective tissue 结缔组织
urinary bladder 膀胱
adipose tissue 脂肪组织
calcium salt 钙盐
spongy bone 海绵骨
blood clotting 血液凝固
muscle tissue 肌肉组织
digestive system 消化系统
reproductive tract 生殖管
cardiac muscle fiber 心肌纤维
intercalated disks 肌间盘
nervous tissue 神经组织
glial cell 神经胶质细胞
cell body 细胞体
LCT Loose Connective Tissue 疏松结缔组织
FCT Fibrous Connective Tissue 纤维结缔组织

4. Notes to the Text A

（1） Adipose tissue has enlarged fibroblasts storing fats and reduced intracellular matrix.

本句中，“storing fats”是一个现在分词短语，做定语，修饰和限定“enlarged fibroblasts”。该短语可以扩展为一个定语从句：“which store fats”。

（2） FCT occurs in tendons, which connect muscle to bone.

本句中，“which connect muscle to bone”是一个非限定性定语从句，对“tendons”做进一步补充说明。

（3） Smooth muscle fibers lack the banding, although actin and myosin still occur.

本句中，“although actin and myosin still occur”是一个让步状语从句，修饰主句的谓语“lack”。“alhough”引导的让步状语从句可以放在句首，也可以放在句尾。主句中不能再用连接词“but”，但可用副词“yet, nevertheless”等。

5. Exercises

（1） Give the word according to the sentence

Sentence	Word
Membranous tissue composed of one or more layers of cells separated by very little intercellular substance and forming the covering of most internal and external surfaces of the body and its organs.	
The portion of the alimentary canal extending from the stomach to the anus and in human beings and other mammals, consisting of two segments, the small intestine and the large intestine.	
A cell that gives rise to connective tissue.	
Generate and separate (a substance) from cells or bodily fluids.	
A tough, elastic, fibrous connective tissue found in various parts of the body, such as the joints, outer ear, and larynx.	
The usually long process of a nerve fiber that generally conducts impulses away from the body of the nerve cell.	
A branched protoplasmic extension of a nerve cell that conduets impulses from adjacent cells inward toward the cell body.	

续表

Sentence	Word
A minute, disklike cytoplasmic body found in the blood plasma of mammals that promotes blood clotting.	
Any of the impulse-conducting cells that constitute the brain, spinal column, and nerves, consisting of a nucleated cell body with one or more dendrites and a single axon.	

(2) Answer the following questions according to the passage

①How many types of tissues are discussed in the text? What are they?

②What are the functions of epithelial tissue?

③How many types of epithelium are there? What are they?

④What are the functions of epithelial cells?

⑤What purposes does connective tissue serve?

⑥What are the two types of connective tissue?

⑦What does cartilage form?

⑧What does muscle tissue do?

⑨How many types of muscle fibers occur in animals? What are they?

⑩What does nervous tissue function?

(3) Translation of the following sentences into Chinese

①The bodies of humans and other mammals contain a cavity divided by the diaphragm into thoracic and abdominal cavities. The body's cells are organized into tissues, which are, in turn, organized into organs and organ systems.

② Epithelial tissues include membranes that cover all body surfaces and glands.

③Connective tissues are characterized by abundant extracellular materials in the matrix between cells. Connective tissue proper may be either loose or dense.

④Skeletal muscles enable the vertebrate body to move. Cardiac muscle powers the heartbeat, while smooth muscles provide a variety of visceral functions.

⑤There are different types of neurons, but all are specialized to receive, produce, and conduct electrical signals.

⑥The vertebrate body is organized into cells, tissues, organs, and organ systems, which are specialized for different functions.

⑦The four primary tissues of the vertebrate adult body-epithelial, connective, muscle, and nerve are derived from three embryonic germ layers.

⑧Smooth muscles are composed of spindle-shaped cells and are found in the organs of the internal environment and in the walls of blood vessels.

⑨Neuroglia are supporting cells with various functions including insulating axons to accelerate an electrical impulse.

⑩Skeletal and cardiac muscles are striated; skeletal muscles, however, are under voluntary control whereas cardiac muscle is involuntary.

课文 A 动物体组织

动物的器官是由不同类型的组织组成的。

上皮组织

上皮组织覆盖身体表面并沿体腔呈线状排列，有支撑功能、保护功能和形成腺体的功能。有 3 种类型的上皮组织：①扁平上皮是扁平状的细胞；②立方上皮是立方体状细胞；③柱状上皮由伸长的细胞组成。

上皮组织分为单层上皮组织和复层上皮组织两大类。单层上皮仅有一层单细胞。复层上皮有一层以上的细胞。假复层上皮有许多单层细胞，但从形状上乍看像两层。

上皮细胞的功能包括：①物质的进出或在体内移动；②保护内环境，抵御外环境；③物质的分泌。

腺体可以是单层上皮细胞，如沿肠道分布的杯状细胞。多细胞腺包括内分泌腺。许多动物的皮肤由上皮组织组成。脊椎动物皮肤内有角蛋白以减少水分流失。还有许多其他动物，如蚯蚓的皮肤分泌黏液或其他物质。

结缔组织

结缔组织在身体内有许多作用，包括：①黏合；②支持；③保护；④形成血液；⑤储存脂肪；⑥填充空间。

非细胞基质将结缔组织细胞彼此分开。这种基质可以是坚硬的（如在骨内）、柔软的（如在疏松结缔组织内）或液体的（如在血液中）。结缔组织分为两类：疏松结缔组织（LCT）和致密结缔组织（FCT）。成纤维细胞被一种纤维基质胶原质分开。胶原质纤维具有弹性和柔性。皮肤和许多内脏器官（如肺、动脉和膀胱）的上皮组织之下有 LCT 组织。这种组织类型也在肌肉、神经和血管上形成保护层。

脂肪组织由变长的能储存脂肪的纤维原细胞和少量的细胞内基质组成。脂肪组织具有能量储藏与绝缘功能。

致密结缔组织内含大量的密集的胶原纤维。在连接肌肉与骨骼的肌腱中有FCT。在关节处连接骨骼与骨骼的韧带也由FCT组成。

软骨与骨骼是“坚硬的”结缔组织。软骨是沉积在细胞间基质内的结构蛋白。它比骨骼软，形成脊椎动物的胚胎骨架及鲨鱼和海星的成年骨架。在人类的耳朵、鼻尖和关节处也有软骨。

骨骼基质中的钙盐增大了骨骼的强度。骨骼也作为身体的钙库（或钙池）。矿物质提供强度，蛋白纤维提供韧性。骨质有两种类型，致密骨质是位于由微管连接的腔隙内的骨细胞。腔隙通常也称作哈弗氏管。海绵状骨质位于骨骼的端部，有被不规则空间分隔开的骨条和骨板。海绵状骨质的固体部分承受压力。

血液是一群被一种液体（血浆）基质分开的结缔组织细胞。血液细胞有两种：红细胞运送氧气。白细胞具有免疫功能。血浆传送分解的葡萄糖、废物、二氧化碳和激素，同时调节血细胞的水平使其平衡。血小板是具有血液凝固作用的细胞碎片。

肌肉组织

肌肉组织通过肌肉细胞（又称肌肉纤维）的收缩促进动物的运动。动物体内有3种类型的肌肉纤维（仅根据肌肉细胞分类）：①骨骼肌；②平滑肌；③心肌。

肌肉纤维是多核的，多个核位于细胞膜之下。这种细胞的大部分被条状、线状的肌原纤维所占据。在每一个肌原纤维内有密集的Z带。肌原纤维节（或肌功能单位）从一个Z带延伸到另一个Z带。每一个肌原纤维节都有粗丝和细丝。粗丝带由肌球蛋白组成，位于肌原纤维节的中心。细丝由肌动蛋白组成，固定在Z带。

骨骼肌（条纹肌）纤维有垂直于细胞长轴的交叉条带。这些细胞的功能与骨骼系统协同引起肌肉随意运动。这些条带是肌动蛋白和肌球蛋白在细胞中的分布。

平滑肌有肌动蛋白和肌球蛋白，但没有条带。这些细胞在非随意运动和（或）自主反应（如呼吸、分泌、射精、分娩及某些反射）中起作用。平滑肌纤维由纺锤状细胞构成。这些肌纤维是消化系统、生殖管和血管的组成成分。

心肌纤维是一种仅仅在心脏中发现的横纹肌。这种细胞呈分叉形状，细胞核通常位于细胞中央。这些细胞通常由肌间盘连接起来。

神经组织

神经组织的功能是对刺激进行分析与综合，以及控制机体对这些刺激做出相应的反应。神经细胞也叫神经元。每一个神经元由一个轴突细胞体和许多树突组成。神经组织主要由两类细胞组成：神经元和神经胶质细胞。神经元传送神经信息。神经胶质细胞与神经元直接接触，并常常包围神经元。

神经元是神经系统的功能单位。人类仅在大脑内就有大约1000亿个神经元。虽然大小和形状各异，但所有的神经元都由三个部分组成。树突从另一个细胞接受信息并传送信息到细胞体。细胞体包含有细胞核、线粒体和真核细胞的其他细胞器。细胞体的信息通常由轴突传出。

Part Two Extensive Reading

1. Text B Some General Features of Animals

Animals are the eaters or consumers of the earth. They are heterotrophs and depend directly or indirectly on plants, photosynthetic protists (algae), or autotrophic bacteria for nourishment. Animals are able to move from place to place in search of food. In most, ingestion of food is followed by digestion in an internal cavity.

Multiicellular Heterotrophs

All animals are multicellular heterotrophs. The unicellular heterotrophic organisms called Protozoa, which were at one time regarded as simple animals, are now considered to be members of the kingdom Protista.

Diverse in Form

Almost all animals (99%) are invertebrates, lacking a backbone. Of the estimated 10 million living animal species, only 42,500 have a backbone and are referred to as vertebrates. Animals are very diverse in form, ranging in size from ones too small to see with the naked eye to enormous whales and giant squids. The animal kingdom includes about 35 phyla, most of which occur in the sea. Far fewer phyla occur in fresh water and fewer still occur on land. Members of three phyla, Arthropoda (spiders and insects), Mollusca (snails), and Chordata (vertebrates), dominate animal life on land.

No Cell Walls

Animal cells are distinct among multicellular organisms because they lack rigid cell walls and are usually quite flexible. The cells of all animals but sponges are organized into structural and functional units called tissues, collections of cells that have joined together and are specialized to perform a specific function; muscles and nerves are tissues types, for example.

Active Movement

The ability of animals to move more rapidly and inmore complex ways than members of other kingdoms is perhaps their most striking characteristic and one that is directly related to the flexibility of their cells and the evolution of nerve and muscle tissues. A remarkable form of movement unique to animals is flying, an ability that is well developed among both insects and vertebrates. Among vertebrates, birds, bats, and pterosaurs (now-extinct flying reptiles) were or are all strong fliers. The only terrestrial vertebrate group never to have had flying representatives is amphibians.

Sexual Reproduction

Most animals reproduce sexually. Animal eggs, which are nonmotile, are much larger than the small, usually flagellated sperm. In animals, cells formed in meiosis function directly as gametes. The haploid cells do not divide by mitosis first, as they do in plants and fungi, but rather fuse directly with each other to form the zygote.

Embryonic Develol Sment

Most animals have a similar pattern of embryonic development. The zygote first undergoes a series of mitotic divisions, called cleavage, and becomes a solid ball of cells, the morula, then a hollow ball of cells, the blastula. In most animals, the blastu1a folds inward at one point to form a hollow sac with an opening at one end called the blastopore. An embryo at this stage is called a gastrula. The subsequent growth and movement of the cells of the gastrula produce the digestive system, also called the gut or intestine. The details of embryonic development differ widely from one phylum of animals to another and often provide important clues to the evolutionary relationships among them.

The Classification of Animals

Two subkingdoms are generally recognized within the kingdom Animalia: Parazoa-animals that for the most part lack a definite symmetry and possess neither tissues nor organs, mostly comprised of the sponges, phylum Porifera; and Eumetazoa-animals that have a definite shape and symmetry and, in most cases, tissues organized into organs and organ systems. Although very different in structure, both types evolved from a common ancestral form and possess the most fundamental animal traits. All eumetazoas form distinct embryonic layers during development that differentiate into the tissues of the adult animal. Eumetazoas of the subgroup Radiata (having radial symmetry) have two 1avers, an outer ectoderm and an inner endoderm, and thus are called diploblastic. All other eumetazoas, the Bilateria (having bilateral symmetry), are triploblastic and produce a third layer, the mesoderm, between the ectoderm and endoderm. No such layers are present in sponges.

2. Notes to the Text B

(1) The unicellular heterotrophic organisms called Protozoa, which were at one time regarded as simple animals, are now considered to be members of the kingdom Protista.

句中“which”引导的非限定定语从句修饰“Protozoa”。非限定定语从句去掉后不影响句子完整性表达。

(2) Of the estimated 10 million living animals species, only 42,500 have a backbone and are referred to as vertebrates.

句中“only”开头的强调句，要用倒装结构。

(3) Animals are very diverse in form, ranging in size from ones too small to see with the naked eye to enormous whales and giant squids.

句中“ranging in…”现在分词短语做状语。分词短语做状语常用于表示方式或伴随情况，说明动作发生的背景或情况。一般情况下，现在分词所表示的动作与谓语所表示的动作同时发生，它的逻辑主语就是句中的主语，谓语动词作为主要动作，而现在分词表示一个陪衬动作，它没有相应的状语从句可以转换，但可以用并列句来转换。过去分词可以说明谓语动作的背景。过去分词与其逻辑主语之间有动宾关系。

句中“too…to…”指“太……而不……”。

Animal cells are distinct among multicellular organisms because they lack rigid cell walls and are usually quite flexible.

句中“because”引导的原因状语从句。

(4) The cells of all animals but sponges are organized into structural and functional units called tissues, collections of cells that have joined together and are specialized to perform a specific function; muscles and nerves are tissues types, for example.

句中“but”指“除了……以外”。“for example”意为用来举例说明，有时可作为独立语，可在句首、句中和句尾，不影响句子其他部分的语法关系。

(5) The ability of animals to move more rapidly and in more complex ways than members of other kingdoms is perhaps their most striking characteristic and one that is directly related to the flexibility of their cells and the evolution of nerve and muscle

tissues.

句中“to move more rapidly and in more complex ways than members of other kingdoms”为动词不定式短语做定语。“that is directly related to the flexibility of their cells and the evolution of nerve and muscle tissues”为定语从句。

(6) The zygote first undergoes a series of mitotic divisions, called cleavage, and becomes a solid ball of cells, the morula, then a hollow ball of cells, the blastula.

句中“called cleavage”为插入语。

3. Answer Questions

(1) What do animals depend directly or indirectly on for nourishment?

(2) What were the unicellular heterotrophic organisms regarded as at one time? What are they now considered?

(3) How many percent of animals are invertebrates?

(4) How many phyla does the animal kingdom include?

(5) Why are animal cells distinct among multicellular organisms?

(6) Why are animals move more rapidly and in more complex ways than members of other kingdoms?

(7) Which is larger, an animal egg or the small flagellated sperm?

(8) How many subkingdoms are generally recognized within the kingdom Animalia?

(9) What are Parazoas?

(10) What are Eumetazoas?

课文B 动物的一般特征

动物是地球生态系统中的消费者。它们是异养生物，直接或间接依赖植物、光合原生生物（藻类），或者自养细菌提供营养。动物能到处寻找食物。在多数情况下，食物的摄取后紧接着的是食物在消化道内的消化。

多细胞异养生物

所有的动物都是多细胞异养生物。单细胞异养生物又叫原生动物，原生动物曾被认为是最简单的动物，现在被认为是原生生物界的成员。

形状的多样性

几乎所有的动物（99%）都是没有脊柱的无脊椎动物。估计1000万种活着的动物中，只有42500种有脊椎，被称为脊椎动物。动物在形状、大小上差别非常大，从小到用肉眼看不到的动物，到巨大的鲸鱼和庞大的鱿鱼。动物界分为35门，大部分生活在海洋中。淡水动物门比海水动物门少得多，陆地动物门更少。节足动物门（蜘蛛和昆虫）、软体动物门（蜗牛）和脊索动物门这三个门中的动物类群是陆地上主要的动物类群。

没有细胞壁

在多细胞机体中，动物细胞比较特殊，因为它们没有坚硬的细胞壁，且通常十分柔软。除海绵外，所有动物细胞都组成组织，组织是机体结构和功能的基本单位。组织是结合在起并履行特殊功能的许多细胞的集合，例如肌肉和神经就是组织类型。

主动运动

动物有比其他界生物以更迅速、以更复杂的方式进行运动的能力，这也许是它们最显著的特征，这与它们细胞的柔软性以及神经与肌肉组织进化有直接关系。动物特有的一种值得注意的运动形式是飞期，它是昆虫和脊椎动物发育良好的一项能力。在脊椎动物中，鸟、蝙蝠和翼龙（已灭绝的会飞的爬行动物）曾经是或现在是强健的飞期者。从来不会飞期的陆地脊椎代表动物只有两栖动物。

有性繁殖

多数动物为有性繁殖。不会运动的动物卵子比小的、带有鞭毛的精子大得多。动物减数分裂中形成的细胞直接具有配子功能。就像它们在植物和真菌中一样，单倍体细胞不会先在有丝分裂中分裂，而是直接相互融合成为受精卵。

胚胎发育

多数动物有一种类似的胚胎发育模式。受精卵首先经受一系列的有丝分裂，即卵裂，变成一个实心细胞球（桑葚胚），然后成为一个有空的细胞球（囊胚）。在多数动物中，囊胚在一点上折叠内陷形成一个一头开口的（称作胚孔）的空囊。这个阶段的胚胎称作原肠胚。接下来，原肠胚细胞的生长和运动产生了消化系统，也叫消化道或肠道。胚胎发育在不同的之间有很大的不同，常常为它们之间进化关系提供线索。

动物的分类

通常认为动物界内有两个亚门：大部分侧生动物身体不对称，无组织和器官，主要由海绵、多孔动物门组成；真后生动物有固定的形状，身体对称，多数情况下，组织形成器官和系统。虽然在结构上很不相同，两种类型都从一个共同的祖先进化而来，并且都具有大部分动物的基本特性。所有真后生动物在发育期间形成独特的胚层，分别分化为成年动物的不同组织。真后生动物的无脊椎辐射动物亚群（具有辐射性对称）有两个胚层，一个外部的外胚层和一个内部的内胚层，这被称为双胚层。所有其他真后生动物、双侧对称（具有左右对称）动物是三胚层动物，即在外胚层和内胚层之间有一个中胚层。而海绵动物则不形成这样的胚层。

Lesson 22

Immunity

Part One Intensive Reading

1. Learning Objective

After learning this lesson, you should understand the following:

(1) Describe the sources of invasion and infection to threat survival of organisms.

(2) Explain what is antigen presentation.

(3) Describe what APCs include.

(4) Define the antibody mediated immunity (or humoral immunity) and explain why it is especially effective at destroying bacteria, extracellular viruses, and other antigens found in body fluids.

(5) Explain why the immune system can recognize millions of antigens with specificity.

(6) Define what is immunologic memory.

2. Text A Immune Response

Among the many threats organisms face are invasion and infection by bacteria, viruses, fungi, and other foreign or disease - causing agents. All organisms have nonspecific defenses (or innate defenses) that provide them with some of the protection they need. This type of defense exists throughout the animal kingdom, from sponges to mammals. Vertebrate animals, however, have an additional line of defense called specific immunity. Specific immunity is also called acquired mmunity, adaptive immunity, or, most simply, an immune response.

Overview

One characteristic of specific immunity is recognition. Immune responses begin when the body recognizes the invader as foreign. This occurs because there are molecules on foreign cells that are different from molecules on the body's cells. Molecules that start immune responses are called antigens. The body does not usually start an immune response against its own antigens because cells that recognize self-antigens are deleted or inactivated. This concept is called self-tolerance and is a key characteristic that defines immune responses.

A second characteristic is specificity. Although all immune responses are similar, each time the body is invaded by a different antigen, the exact response is specific to that antigen. For example, infection with a virus that causes the common cold triggers a response by a different set of cells than infection with bacteria that causes *strep* throat.

A third characteristic is memory. After an antigen is cleared from the body, immunological memory allows an antigen to be recognized and removed more quickly if encountered again.

Antigen Presentation

Three groups of white blood cells are involved in starting an immune response. Although immune responses can occur anywhere in the body these cells are found, they primarily occur in the lymph nodes and spleen. These organs contain large numbers of antigen-presenting cells (APCs), T lymphocytes (or T cells), and B lymphocytes (or B cells).

APCs include macrophages, dendritic cells, and B cells. These cells encounter the foreign invader and present the invader's antigens to a group of T cells called helper T cells (TH cells). APCs engulf an invader and bring it inside the cells. The APC then breaks the invader apart into its antigens and moves these antigens to its cell surface. Receptors are cell surface proteins that can attach to antigens. Each TH cells has a different receptor, allowing each cell to recognize a different antigen.

The APC "shows" the antigen to the TH cells until there is a match between a TH cell receptor and the antigen. The contact between the two cells stimulates the TH cell to divide rapidly. This process is called clonal selection because only the TH cells that recognize the foreign invader are selected to reproduce. Stimulated TH cells also produce chemical messengers called cytokines. Cytokines are made by all immune cells and control the immune response.

Antigen Clearance

The large numbers of TH cells activate two other populations of white blood cells: cytotoxic T cells (TC cells) and B cells. Like TH cells, each TC cell and B cell has receptors that match one antigen. This is why the immune system can recognize millions of antigens with specificity. The cells with the appropriate receptor encounter the antigen, preparing them for activation. They receive the final signal necessary for clonal selection from TH cells and cytokines. Cloned TC cells attach to invaders they recognize and release a variety of chemicals that destroy the foreign cell. Because this must happen through cell-to-cell contact, it is called cell-mediated immunity (or cellular immunity). It is especially effective at destroying abnormal body cells, such as cancerous cells or virus-infected cells. Cloned B cells destroy foreign invaders differently. After activation by TH cells, B cells release proteins called antibodies. Antibodies travel through the body's fluids and attach to antigens, argeting them for destruction by nonspecific defenses. This type of immune response is called antibodymediated immunity (or humoral immunity). It is especially effective at destroying bacteria, extracellular viruses, and other antigens found in body fluids.

Immunologic Memory

A primary immune response happens the first time that the body encounters a specific antigen. It takes several days to begin and one or two weeks to reach maximum activity. A secondary immune response occurs if the body encounters the same antigen at a 1ater time. It takes only hours to begin and may peak within a few days. The invader is usually removed before it has a chance to cause disease. This is because some of the cloned TC cells and B cells produced during a primary immune response develop into memory cells. These cells immediately become activated if the antigen appears again. The complex interactions among cells described above are not necessary.

In fact, this is what happens when an individual is immunized against a disease. The vaccination (using weakened or killed pathogens) causes a primary immune response (but not the disease) and the production of memory cells that will provide protection if exposed to the disease-causing agent.

Immune System Disorders

Studying immune responses also allows scientists to understand immune system diseases. For example, hypersensitivity disorders occur when the immune system overreacts to an antigen, causing damage to healthy tissues. The result of this excessive antibody and TC cells activity can be relatively harmless (as with allergies to pollen, poison ivy, or molds) or deadly (as with autoimmune diseases or allergies to bee Venom and antibiotics). At the opposite end of the spectrum are immunodeficiency diseases, conditions in which the body does not respond effectively against foreign invaders. HIV (human immunodeficiency virus) infection causes AIDS (acquired immunodeficiency syndrome) by attacking TH cells. Occasionally an individual is born with a deficient immune system, but these disorders are usually acquired (for example, from radiation treatment, chemotherapy, or infection with HIV). Whatever the cause, the individual has a more difficult time fighting infections.

3. New Words and Phrases

immune [ɪˈmjuːn] *adj.* 免疫的
invader [ɪnˈveɪdər] *n.* 入侵物，入侵者
Streptococcus *n.* 链球菌
specificity [ˌspesɪˈfɪsətɪ] *n.* 特异性
inactivate [ɪnˈæktɪveɪt] *vt.* 使不活动；使变不活泼；去激活
spleen [spliːn] *n.* 脾
lymphocyte [ˈlɪmfəʊsaɪt] *n.* 淋巴细胞
macrophage [ˈmækrəʊfeɪdʒ] *n.* 巨噬细胞
engulf [ɪnˈgʌlf] *vt.* 吞没；吞食
dendritic [denˈdritik] *adj.* 树枝状的
cytokine [ˈsaɪtəkaɪn] *n.* 细胞因子，细胞活素类物质
cytotoxic [saɪtəˈtɑksɪk] *adj.* 细胞毒素的
vaccination [ˌvæksɪˈneɪʃən] *n.* 接种疫苗
hypersensitivity [ˌhaɪpərˌsɛnsəˈtɪvɪti] *n.* 超敏性
allergy [ˈælərdʒi] *n.* 过敏
venom [ˈvenəm] *n.* 毒液
immunodeficiency [ˌɪmjʊnəʊdɪˈfɪʃənsɪ] *n.* 免疫缺陷
autoimmune [ˌɔːtoʊɪˈmjuːn] *adj.* 自体免疫的
chemotherapy [ˌkiːmoʊˈθerəpi] *n.* 化学疗法，化疗
immune response 免疫反应
immune system 免疫系统
nonspecific defense 非特异性防疫系统
specific immunity 特异性免疫
acquired immunity 获得性免疫
adaptive immunity 适应性免疫
strep throat 脓毒性咽喉炎
antigen presentation 抗原呈递

lymph node 淋巴结
clonal selection 克隆选择
antigen clearance 抗原清除
immunologic memory 免疫记忆
humoral immunity 体液免疫
bee venom 蜂毒
APCs antigen-presenting cells 抗原呈递细胞
AIDS acquired immunodeficiency syndrome 艾滋病，获得性免疫缺陷综合征

4. Notes to the Text A

(1) This occurs because there are molecules on foreign cells that are different from molecules on the body's cells.

本句中，"because there are molecules on foreign cells that are different from molecules on the body's cells"是一个原因状语从句，修饰谓语"occurs"。在该状语从句中，"that are different from molecules on the body's cells"是一个定语从句，修饰和限定第一个"molecules"。

(2) Although all immune responses are similar, each time the body is invaded by a different antigen, the exact response is specific to that antigen.

本句中，"Although all immune responses are similar"是一个让步状语从句，"each time the body is invaded by a different antigen"是一个时间状语从句，它们都修饰谓语"is specific to that antigen"。

(3) Cloned TC cells attach to invaders they recognize and release a variety of chemicals that destroy the foreign cell.

本句中，"they recognize"是一个省略了"that"的定语从句，修饰和限定"invaders"。"that destroy the foreign cell"也是一个定语从句，修饰和限定"a variety of chemicals"。

(4) At the opposite end of the spectrum are immunodeficiency diseases, conditions in which the body does not respond effectively against foreign invaders.

为了保持句子的平衡，本句使用了倒装句。本句中，主语是"immunodeficiency diseases"，"conditions in which the body does not respond effectively against foreign invaders"是对"immunodeficiency diseases"的补充说明，"in which the body does not respond effectively against foreign invaders"是一个定语从句，修饰和限定"conditions"。"are"是系动词，"At the opposite end of the spectrum"为表语。

5. Exercises

（1） Give the word according to the sentence

Sentence	Word
Protected against a particular disease by particular substances in the blood.	
Of，relating to，or producing a toxic effect on cells.	
An innate，acquired，or induced inability to develop a normal immune response.	
A large，highly vascular lymphoid organ，lying in the human body to the left of the stomach below the diaphragm，serving to store blood，disintegrate old blood cells，filter foreign substances from the blood，and produce lymphocytes.	
Make sth. not work or operate any more，make inactive.	
To put or force in inappropriately，especially without invitation，fitness，or permission.	
The chemical messengers which are made by all immune cells and control the immune response.	
The treatment of disease using chemical agents or drugs that are selectively toxic to the causative agent of the disease，such as a virus，bacterium，or other microorganism.	
A poisonous secretion of an animal，such as a snake，spider，or scorpion，usually transmitted by a bite or sting.	
An abnormally high sensitivity to certain substances，such as pollens，foods，or microorganisms.	

（2） Answer the following questions according to the passage

①What are the threats mentioned in the text organisms face?

②What are the three characteristics of immune response?

③What are molecules that start immune responses called?

④Why doesn ’ t the body usually start an immune response against its own antigens?

⑤After an antigen is cleared from the body，if encountered again，what does immune-logical memory do?

⑥What do APCs include?

⑦What are cytokines made? And what do they do?

⑧What does antibodymediated immunity refer to?

⑨The invader is usually removed before it has a chance to cause disease. Why?

⑩What does HIV infection cause, and how?

(3) Translation of the following sentences into Chinese

①The surface defenses of the body consist of the skin and the mucous membranes lining the digestive and respiratory tracts, which eliminate many microorganisms before they can invade the body tissues.

②Inflammation aids the fight against infection by increasing blood flow to the site and raising temperature to retard bacterial growth.

③The inflammatory response aids the mobilization of defensive cells at infected sites.

④Nonspecmc defenses include physical barriers such as the skin, phagocytic cells, killer cells, and complement proteins.

⑤The immune system evolved in animals from a strictly nonspecific immune response in invertebrates to the two-part immune defense found in mammals.

⑥Skin not only defends the body by providing a nearly impenetrable barrier, but also reinforces this defense with chemical weapons on the surface.

⑦Physical and anatomic barriers that tend to prevent the entry of pathogens are an organism's first line of defense against infection.

⑧Innate immunity is not specific to any one pathogen but rather constitutes a first line of defense, which includes anatomic, physiologic, endocytic and phagocytic, and inflammatory barriers.

⑨Adaptive immune responses exhibit four immunologic attributes: specificity, diversity, memory, and self/nonself recognition.

⑩Immunity is the state of protection against foreign organisms or substances (anti-gens). Vertebrates have two types of immunity, innate and adaptive.

课文A 免疫反应

在生物体所面对的威胁中，有一部分是由细菌、病毒、真菌和其他外来的致病因子所引起的。所有的生物体都有非特异性免疫系统（或先天免疫系统），可以提供生物所需的部分保护功能。这种防御系统存在于从海绵动物到哺乳动物的整个动物界中。但是在脊椎动物中还有一种防御系统，称之为特异性免疫，这种免疫机制也被称为获得性免疫，或者简称为免疫反应。

综述

特异性免疫的一个特点是具有识别功能。当生物体识别出外来入侵物时，免疫反应就开始了，原因是外来细胞和自身体细胞具有不同的分子。能引起免疫反应的分子被称之为抗原。生物体通常不会对自身的抗原产生免疫反应，因为能够识别自身抗原的细胞已被去除或灭活。这种机制被称为自身耐受，它是定义免疫反应的一个关键特征。

第二个特征是特异性。尽管所有的免疫反应都比较相似，但是当机体受到不同的抗原入侵时，会产生与这种抗原相对应的免疫反应，如与脓毒性咽喉炎细菌引起的感染相比较，感冒病毒感染会引发不同种类细胞的反应。

第三个特点是具有记忆性。在抗原被从身体中清除以后，如果再遇到这种抗原，免疫记忆就会识别这种抗原，并能很快地将其清除出去。

抗原呈递

三种白细胞与免疫反应的启动有关。尽管免疫反应可在身体任何一个具有这些细胞的部位发生，但最初只能在淋巴结和脾中发生。这些器官含有大量的抗原呈递细胞（APCs）、T 淋巴细胞（或 T 细胞）和 B 淋巴细胞（或 B 细胞）。

抗原呈递细胞包括巨噬细胞、树突细胞和 B 细胞。这些细胞遇到外来入侵物之后，把外来入侵物抗原呈递给辅助性 T 细胞（TH 细胞）。抗原呈递细胞再把外来物体吞入，并把它带入细胞内。然后抗原呈递细胞把入侵物分解成抗原，再把这些抗原转移到细胞表面。受体是能够与抗原结合的细胞表面蛋白，每一个 TH 细胞都有不同的受体，这就使得不同的细胞只能识别特定的抗原。

抗原呈递细胞把抗原呈递给 TH 细胞，直到 TH 细胞受体与抗原相配套为止。两个细胞之间的接触会刺激 TH 细胞加速分裂，这个过程称之为克隆选择，因为只有识别了外来入侵物的 TH 细胞才会进行增殖。受到刺激的 TH 细胞也会产生一些称为细胞因子的化学信息素。所有的免疫细胞都会产生细胞因子，来调控免疫反应。

抗原清除

大量的辅助性 T 细胞激活了其他两种白细胞：细胞毒性 T 细胞（TC 细胞）和 B 细胞。与 TH 细胞相似，每一个 TC 细胞和 B 细胞都有与抗原相配套的受体，这就是为什么免疫系统能根据特异性识别几百万种抗原的原因。当具有配套受体的细胞遇到抗原时，就做好了激活的准备。它们收到从 TH 细胞和细胞因子发出的克隆选择所需的最终信号。克隆的 TC 细胞附着在它们识别出来的入侵物上，然后释放出多种化学物质来消灭外来细胞。由于必须要通过细胞与细胞的接触才能发生反应，因此这个反应就被称为细胞媒介免疫（或细胞免疫）。这对于消灭不正常的细胞非常有效，如癌细胞或病毒感染细胞。克隆的B细胞通过不同的方式来消灭入侵物。在TH细胞激活之后，B细胞释放一种称为抗体的蛋白。抗体通过体液流至全身各处，再与抗原结合，利用非特异性免疫机制消灭目标细胞。这种免疫反应我们称为抗体媒介免疫（或体液免疫）。这种机制对于清除细菌、胞外病毒和其他存在于体液中的抗原特别有效。

免疫记忆

当机体首次遇到一种特异性抗体时，最初的免疫反应就发生了。启动反应需要几天时间，然后用1~2周的时间来达到最大活性。如果以后机体再遇到同样的抗原，就会发生第二次免疫反应，这次只需要几个小时来启动免疫反应，而且可能在几天之内就会达到免疫反应的最大活性。这样外来入侵物在引起疾病之前就会被去除掉。这是因为在最初免疫反应发生时产生的一些克隆的TC细胞和B细胞发展成为了记忆细胞。如果同样的抗原出现，那么这些细胞就会立即被激活。前述的细胞间复杂的相互作用将不再叙述。

实际上，当一个生物体对某种疾病产生免疫力时，就会发生这样的反应。疫苗（利用弱毒或灭活病原体）首先引起最初的免疫反应（但不是疾病），然后产生的记忆细胞将会在遇到病原体时提供保护作用。

免疫系统失调

通过对免疫反应进行研究，科学家了解了免疫系统疾病，例如，超敏反应是指机体免疫系统对某种抗原所表现的异常增高的免疫应答，可损害健康组织。过多的抗体和TC细胞活动的结果可以是无害的（如对花粉、毒葛或霉菌过敏），也可能是致命的（如自身免疫系统疾病或对蜂毒和抗生素过敏）。与上述相反的另一个极端是免疫缺陷疾病，这种情况下机体不能有效地对入侵物产生免疫反应。HIV（人类免疫缺陷病毒）攻击TH细胞，其感染引起AIDS（获得性免疫缺陷综合征）。有时一些个体在出生时就有免疫缺陷，但是这些异常通常是后天获得的（例如由于放射线辐射、化疗和HIV感染）。不管是什么原因，具有免疫缺陷的个体将很难抵抗感染。

Part Two Extensive Reading

1. Text B Cellular Counterattack

The surface defenses of the vertebrate body are very effective but are occasionally breached, allowing invaders to enter the body. At this point, the body uses a host of nonspecific cellular and chemical devices to defend itself. We refer to this as the second line of defense. These devices all have one property in common: they respond to any microbial infection without pausing to determine the invader's identity.

Although these cells and chemicals of the nonspecific immune response roam through the body, there is a central location for the collection and distribution of the cells of the immune system; it is called the lymphatic system. The lymphatic system consists of a network of lymphatic capillaries, ducts, nodes and lymphatic organs, and although it has other functions involved with circulation, it also stores cells and other agents used in the immune response. These cells are distributed throughout the body to fight infections, and also stored in the lymph nodes where foreign invaders can be eliminated as body fluids pass through.

Cells that Kill Invading Microbes

Perhaps the most important of the vertebrate body's nonspecific defenses are white blood cells called leukocytes that circulate through the body and attack invading microbes within tissues. There are three basic kinds of these cells, and each kills invading microorganisms differently.

Macrophages ("big eaters") are large, irregularly shaped cells that kill microbes by ingesting them through phagocytosis, much as an amoeba ingests a food particle. Within the macrophage, the membrane-bound vacuole containing the bacterium fuses with a lysosome. Fusion activates lysosomal enzymes that kill the microbe by liberating large quantities of oxygen free-radicals. Macrophages also engulf viruses, cellular debris, and dust particles in the lungs. Macrophages circulate continuously in the extracellular fluid, and their phagocytic actions supplement those of the specialized phagocytic cells that are part of the structure of the liver, spleen, and bone marrow. In response to an infection, monocytes (an undifferentiated leukocyte) found in the blood squeeze through capillaries to enter the connective tissues. There, at the site of the infection, the monocytes are transformed into additional macrophages.

Neutrophils are leukocytes that, like macrophages, ingest and kill bacteria by phagocytosis. In addition, neutrophils release chemicals (some of which are identical to household bleach) that kill other bacteria in the neighborhood as well as neutrophils themselves.

Natural killer cells do not attack invading microbes directly. Instead, they kill cells of the body that have been infected with viruses. They kill not by phagocytosis, but rather by creating a hole in the plasma membrane of the target cell. Proteins, called perforins. are released from the natural killer cells and inserted into the membrane of the target cell, forming a pore. This pore allows water to rush into the target cell, which then swells and bursts. Natural killer cells also attack cancer cells, often before the cancer cells have had a chance to develop into a detectable tumor. The vigilant surveillance by natural killer cells is one of the body' most potent defenses against cancer.

Proteins that Kill Invading Microbes

The cellular defenses of vertebrates are enhanced by a very effective chemical defense called the complement system. This system consists of approximately 20 different proteins that circulate freely in the blood plasma. When they encounter a bacterial or fungal cell wall, these proteins aggregate to form a membrane attack complex that inserts itself into the foreign cell's plasma membrane, forming a pore like that produced by natural killer cells. Water enters the foreign cell through this pore, causing the cell to swell and burst. Aggregation of the complement proteins is also triggered by the binding of antibodies to invading microbes.

The proteins of the complement system can augment the effects of other body defenses. Some amplify the inflammatory response by stimulating histamine release; others attract phagocytes to the area of infection; and still others coat invading microbes, roughening the microbes' surfaces so that phagocytes may attach to them more readily.

Another class of proteins that play a key role in body defense are interferons. There are three major categories of interferons: alpha, beta, and gamma. Almost all cells in the body make alpha and beta interferons. These polypeptides act as messengers that protect normal cells in the vicinity of infected cells from becoming infected. Though viruses are still able to penetrate the neighboring cells, the alpha and beta interferons prevent viral replication and protein assembly in these cells. Gamma interferon is produced only by particular lymphocytes and natural killer cells. The secretion of gamma interferon by these cells is part of the immunological defense against infection and cancer.

2. Notes to the Text B

(1) The lymphatic system consists of a network of lymphatic capillaries, ducts, nodes and lymphatic organs, and although it has other functions involved with circulation, it also stores cells and other agents used in the immune response.

句中 "and" 引导的并列句，后一个分句包含一个由 "although" 引导的状语从句。

(2) Perhaps the most important of the vertebrate body's nonspecific defenses are white blood cells called leukocytes that circulate through the body and attack invading microbes within tissues.

句中 "called leukocytes" 分词短语做定语修饰 "cells"。"that" 引导定语从句修饰 "leukocytes"。

(3) Macrophages ("big eaters") are large, irregularly shaped cells that kill microbes by ingesting them through phagocytosis, much as an amoeba ingests a food particle.

句中 "irregularly shaped cells that kill microbes by ingesting them through phagocytosis" 分词短语做插入语，that 引导定语从句修饰 "cells"。

(4) Fusion activates lysosomal enzymes that kill the microbe by liberating large quantities of oxygen free-radicals.

句中 "that" 引导的定语从句，主句是 "Fusion activates lysosomal enzymes"。

(5) Though viruses are still able to penetrate the neighboring cells, the alpha and beta interferons prevent viral replication and protein assembly in these cells.

句中 "Though" 引导让步状语从句。

3. Answer Questions

(1) What is the second line of defense?

(2) What property do nonspecific cellular and chemical devices have in common?

(3) What is the lymphatic system?

(4) What are the most important of the vertebrate body's nonspecific defenses? What do they do?

(5) What are macrophages?

(6) What are neutrophils?
(7) What's the functions of natural killer cells?
(8) How to enhance the cellular defenses of vertebrates?
(9) What can the proteins of the complement system do?
(10) How many major categories of interferons are there? What are they?

课文B 细胞反击

脊椎动物的体表防线是非常有效的，但是有时候也会被攻破，使得入侵物能进入到机体内部。在这种情况下，机体会运用一种非特异性的细胞和化学装置来进行防御。我们称之为第二道防线。这种装置都有一个共同的特性：不需要确定入侵物的类型，对任何微生物的感染都做出反应。

尽管非特异性免疫反应的细胞和化学物质会分散到全身各处，但是还有一个免疫中心来收集和分配免疫系统中的细胞，这个中心就是淋巴系统。淋巴系统包括一个由淋巴毛细管、淋巴管、淋巴结和淋巴器官组成的网络。尽管淋巴系统还具有与循环有关的其他功能，但它也能存储免疫反应所需的细胞和其他介质。这些细胞会被分配到全身各处以抗感染，这些细胞也存储在淋巴结上，在这里当体液通过时外来入侵物就会被清除。

杀灭入侵微生物的细胞

脊椎动物机体中最重要的非特异性免疫物质是白细胞，这种细胞通过血液循环到全身各处，然后攻击侵入体组织的微生物。白细胞有三种基本类型，每一种都可以杀灭不同的入侵微生物。

巨噬细胞是一种体积较大、形状不规则的细胞，通过噬菌作用把微生物吞入胞内，然后杀死微生物，与变形虫吞噬食物颗粒非常相似。在巨噬细胞内，由膜包被的液泡中含有溶酶体与细菌的融合体。这种融合激活了溶酶体中酶的活性，这些酶可通过释放大量的氧自由基来杀灭微生物。巨噬细胞也会吞噬病毒、细胞碎片和肺中的灰尘颗粒。巨噬细胞在细胞外液中不断地流动，巨噬细胞的噬菌作用是对肝、脾和骨髓结构一部分的特殊的噬菌细胞的作用的补充。对感染做出反应的过程中，血液中的单核细胞（未分化的白细胞）通过毛细血管进入结缔组织。在感染部位，单核细胞转化成巨噬细胞。

嗜中性粒细胞是一种与巨噬细胞相似的白细胞，通过噬菌作用吞入并杀灭微生物。另外，嗜中性粒细胞释放一些化学物质（有些与家用漂白剂相似），杀灭邻近的微生物和嗜中性粒细胞自身。

自然杀伤细胞不会直接攻击入侵的微生物，它们会杀灭机体中受病毒感染的细胞。它们不是通过吞噬作用，而是在目标细胞的细胞膜上打一个孔。自然杀伤细胞释放一种被称为穿孔素的蛋白质，插入到目标细胞膜中，形成一个孔。产生的孔会使得水分涌入目标细胞中，目标细胞膨胀然后破裂。自然杀伤细胞也会攻击癌细胞，一般是在癌细胞有机会发展成为可检测到的肿瘤之前。自然杀伤细胞灵敏的监测机制是身体最有效的防癌机制之一。

杀灭入侵微生物的蛋白质

一种称之为补体系统的化学防御机制有效地增强了脊椎动物的细胞防御能力。这个系统包括20来种在血浆中自由循环的蛋白质。当它们遇到细菌或真菌细胞壁时，这些蛋白质集合成膜攻击复合体，插入到外来细胞的质膜中，形成一个与杀伤细胞作用相类似的孔，水分再由这个孔进入到细胞内，引起细胞膨胀和破裂。抗体结合到入侵的微生物中也会激发起补体蛋白的凝集。

补体系统的蛋白质能增强身体其他防御系统的作用。有些补体通过组胺的释放增强炎症反应；其他补体会吸引巨噬细胞赶赴受感染位点；还有一些能包在入侵的微生物表面，使其表面变粗糙，从而有利于吞噬细胞黏附在它们表面。

另有一些与身体防御有关的蛋白质是干扰素。有三种主要的干扰素：α，β和γ干扰素。机体中几乎所有的细胞都产生α，β干扰素。这些多肽起着信息素的作用，能保护邻近受感染细胞的正常细胞免受感染。尽管病毒仍然能够穿透邻近细胞，但α，β干扰素能够防止病毒在这些细胞中的复制和病毒蛋白质的装配。γ干扰素只由特殊的淋巴细胞和自然杀伤细胞生成。这些细胞中γ干扰素的分泌是免疫防御中抗感染和抗癌机制的组成部分。

Lesson 23
Pathology

Part One Intensive Reading

1. Learning Objective

After learning this lesson, you should understand the following:

(1) Explain what pathology and its study are.

(2) Describe the main steps that are involved in the evolution of a clinician.

(3) Define clinical sign, lesion, etiology, pathogenesis, diagnosis and prognosis.

(4) Explain what means to be sick.

(5) Describe the exhibits clinical signs when animals are observed as diarrhea.

(6) Explain why the study of pathology involves a new language.

2. Text A Study on Pathology

Pathology is the science of the study of disease. It investigates the essential nature of disease and is usually summarized as the study of the functional and morphological changes in the tissues and fluids of the body during disease.

The end product of a medical curriculum is a clinician whose function is the diagnosis and treatment of disease. Several main steps are involved in the evolution of a clinician. It is necessary to learn the normal development, structure and function of the body, and this occurs through the media of embryology, anatomy, histology, genetics, pharmacology, biochemistry and physiology. This step is a prelude to learning what goes wrong with the machine that is called the body. The development of structural and functional changes in cells, tissues, fluids and organs that result in malfunction and disease is studied in pathology. The agents that cause disease become known in the study of microbiology, parasitology and toxicology. With the background of knowing what goes wrong and how, the next step for the prospective clinician is to learn how to find out what the problem is in the body of a sick individual and how to treat it by medical or surgical means. Better still, after all this study of disease comes the revelation that it would be easier in the long run to prevent the disease, if possible, rather than to find and treat it. Of these steps in the development of a clinician, pathology is a cornerstone, because the clinician cannot rationally diagnose and treat without understanding the disease process with which he is dealing.

What does it mean to be sick? If an individual is sick, it is assumed that some part of the body is not functioning properly, and it is expected that with proper diagnostic procedures the problem may be found and treated. What of the individual who was "perfectly healthy" but who died suddenly of a stroke or heart attack? There must have been in a serious state of disease in some tissues even though the individual had never been "clinically" sick. In fact, some of the cells and tissues in the body of the normal individual who dies suddenly were sick and were not functioning properly. Pathology is the study of the molecular, biochemical, functional and morphological aspects of disease in the fluids, cells, tissues and organs of the body. Changes brought on by these aspects may load to clinical illness or may be well developed but not apparent to the individual with the problem.

A number of specific terms are used in pathology and medicine which require explanation. Some of the most important are: clinical sign, lesion, etiology, pathogenesis, diagnosis and prognosis.

Suppose an individual has diarrhea. The feeling of abdominal discomfort is a clinical symptom that can be described by humans but not by animals. The animal exhibits clinical signs that are observed as diarrhea, not eating, an anxious expression and perhaps dehydration. There must be a functional derangement in the fluid transport at the cellular level in the intestinal mucosa. It may or may not be visible as a microscopic abnormality in the cell, but there is a functional problem. This problem is called a lesion and may be either functional or morphological or both. The lesion is the "abnormality" in the tissue. This word is perhaps the most commonly used word in pathology. Aetiology refers to the cause of the diarrhea and might be bacteria or virus, some unusual food or a sudden change in diet.

Pathology is the study of what and how happened and the lesion is what. The sequence of events from the point at which the lesion began through its entire development is called the pathogenesis. It is necessary to know the pathogenesis of lesions in order to make a rational judgment for treatment, control and prevention of disease. Pathogenesis is how the step by step progression from the normal state through to the abnormal structural or functional state. The usual sequence is to find a lesion, to identify it and then attempt to determine the pathogenesis by investigation of the circumstances and the sequence of events that lead to the lesion.

The study of pathology involves a new language. A primary objective of that study is to master the terminology by learning the definitions used and limitations of the language of pathology in its role in the description of lesions and their pathogenesis and etiology. This is the theory of pathology but not the practice. The practice involves being able to describe lesions, to recognize the disease process and to explain how it might have occurred. This requires practical experience, exposure to specimens and problem - solving ability. The theory is dangerous without the practical ability. Depending on the objective for studying pathology, there may be good reason for just learning the words and having a general idea about specific disease processes for general interest. This may be a rewarding and satisfying experience. A license to practice, however, implies knowledge and the ability to apply it. Technical training is essential to carry out diagnostic procedures, but knowledge and practical experience are restricted to interpret, diagnose and provide an accurate prognosis.

3. New Words and Phrases

abdominal [æbˈdɑːmɪnl] *adj.* 腹部的
etiology [ˌiːtɪˈɒlədʒɪ] *n.* 病因学
intestinal [ɪnteˈstaɪnəl] *adj.* 肠的
mucosa [mjuːˈkəʊsə] *n.* 黏膜
lesion [ˈliːʒən] *n.* 损害，病变
parasitology [ˌpærəsɪˈtɒlədʒɪ] *n.* 寄生虫学
pathology [pəˈθɒlədʒɪ] *n.* 病理学
prognosis [prɒgˈnəʊsɪs] *n.* 预断病情；预后；预测

4. Notes to the Text A

A primary objective of that study is to master the terminology by learning the definitions used and limitations of the language of pathology in its role in the description of lesions and their pathogenesis and etiology.

这是一个简单句，但句子较长。主语为“objective”，谓语是“is”，宾语是“to master terminology”，其后的“by+动词 ing”做状语。

译文如下：该研究的主要目标是通过描述（疾病的）病变、发病机制和病因学，来学习病理学语言的概念及其用途和范围，从而掌握这些术语。

5. Exercises

(1) Fill in the blanks by finishing the sentences according to the passage

① Pathology investigates the essential nature of disease and is usually summarized as the study of the functional and morphological changes in the tissues and fluids of the body during disease. In this sentence, "essential nature" can be replaced by ________.

②Pathogenesis is how the step by step progression from the normal state through to the abnormal structural or ________ state.

③The end product of a medical curriculum is a ________ whose function is diagnosis and treatment of disease.

④It is necessary to learn the normal development, structure and function of the body, and this occurs through the media of embryology, ________, histology, genetics, pharmacology, biochemistry and physiology.

⑤ Pathology is the study of the molecular, biochemical, functional and ________ aspects of disease in the fluids, cells, tissues and organs of the body.

(2) Answer the following questions according to the passage

①Does pathology have its own terminological system?

②Is it true that pathology is the study of the molecular, biochemical, functional and morphological aspects of disease in the fluids, cells, tissues and organs of the body?

③What is pathogenesis?

④According to the text, what is the difference between the theory and practice of pathology?

⑤What is essential to carrying out diagnostic procedures?

(3) Translation of the following sentences into Chinese

①Pathology is the science of the study of disease. It investigates the essential nature of disease and is usually summarized as the study of the functional and morphological changes in the tissues and fluids of the body during disease.

②If an individual is sick, it is assumed that some part of the body is not functioning properly, and it is expected that with proper diagnostic procedures the problem may be found and treated.

③A number of specific terms are used in pathology and medicine which require explanation.

④The feeling of abdominal discomfort is a clinical symptom that can be described by humans but not by animals.

⑤Aetiology refers to the cause of the diarrhea and might be bacteria or virus, some unusual food or a sudden change in diet.

⑥Pathogenesis is how the step by step progression from the normal state through to the abnormal structural or functional state.

⑦A primary objective of that study is to master the terminology by learning the definitions used and limitations of the language of pathology in its role in the description of lesions and their pathogenesis and etiology.

课文A 病理学研究

病理学是研究疾病的一门科学。它研究疾病的基本特征，并通常将它归结为研究患病过程中机体组织和体液在功能上和形态上的变化。

医学教育的最终目的是培养临床医师，而临床医师的作用是诊断和治疗疾病。一个临床医师的成长、成熟包括好多重要步骤。其中学习机体的正常发育、结构和功能是必要的，这部分知识可以通过学习胚胎学、解剖学、组织学、遗传学、药理学、生物化学和生理学而获得。这只是学习机体哪里出错的第一步，还要学习病理学，研究导致功能障碍和疾病的细胞、组织、体液以及器官的结构和功能的改变。学习微生物学、寄生虫学和毒理学以知道引发疾病的原因。有了哪里发病和怎样发病的背景知识，未来的临床医师下一步是学习如何找到患者的问题所在，以及如何通过内科和外科方法进行治疗。对于疾病的发生、发展了解之后，最好从长远考虑，如果可能应是防病重于治病。在培养临床医师的这些步骤中，病理学是基础，因为如果不了解所治疾病的病程，他就不能合理地诊断和治疗疾病。

患病是什么意思？如果一个人患病了，这说明其身体的某个部位不能正常地发挥功能，同时希望利用恰当的诊断方法来发现问题并进行治疗。为什么一个看起来“相当健康”的人会突然死于中风和心脏病呢？这是因为在他的身体的某些组织必定发生了严重病变，尽管他未表现出临床症状。事实上，对于那些猝死的人来说，其体内的某些细胞和组织已经不能正常工作了。病理学就是从分子、生物化学、功能和形态等方面研究机体体液、细胞、组织和器官的病变。这些方面的改变会导致临床上的疾病，或者病变进一步发展，而无明显临床症状。

在病理学和医学中有许多专业术语，需要解释一下。最重要的有临床症状、病变、病因学、发病机理、诊断和预后。

假设某人得了腹泻，腹部不舒适的感觉是一种临床症状，人类可以描述，但动物不能描述。动物表现出的临床症状是能够观察到的腹泻、拒食、不安，或许脱水。动物小肠黏膜细胞层一定有体液运输的功能紊乱。细胞的异常或许不能用显微镜观察到，但确实存在功能上的问题，这种现象就称作病变，这种病变可能是功能上的，也可能是形态上的，或者二者兼有。病变是组织中的“异常变化”。这个词可能是病理学中最常用的名词。病因学是指引起腹泻的原因，可能是细菌性的，也可能是病毒性的，还可能是一些异常的食物或饲料的突然变更。

病理学是研究发生了什么和怎样发生的，病变的严重程度。从病变开始引发的一系列症状到随后的整个发病过程，称作发病机制。要想找到正确的治疗方法，有效地控制和预防疾病，就应该了解病变的发病机制。发病机制就是指疾病是怎样从正常状态一步步发展到结构和功能异常的过程。通常的过程是找到病变并识别病变，然后试图通过对引起病变的环境和结果的调查来确定病变的发病机制。

研究病理学涉及一种“新语言”。该研究的主要目标是通过描述（疾病的）病变、发病机制和病因学，来学习病理学语言的概念及其用途和范围，从而掌握这些术语。这是病理学的理论，而不是实践。实践包括能够描述病变，辨别疾病过程以及解释疾病是如何发生的。这需要有实践经验、接触病例和有解决问题的能力。只有理论而无实践能力是很危险的。若仅依据研究病理学的目标，有理由单单学习这些术语，对一些特殊的病理过程有一个总体的印象，可能就会令人满足。然而，要获得执照做医师，不仅要有知识，还要有能力应用知识。要完成诊断程序，必须有技术培训；解释、诊断和给出确切的预后需要知识和实践经验。

Part Two Extensive Reading

1. Text B Cardiomyopathy in Ferrets

Cardiomyopathy is a common disease in the American lines of ferrets, which has a presumed genetic basis. Several forms of this condition may be seen - dilatative, hypertrophic, and a restrictive form in which there is marked replacement of myocardium by fibrous connective tissue, with minimal change in chamber area. Signs of cardiomyopathy may be seen as early as 1 year of age in severely affected animals, but are more common between 5 and 7 years of age.

Grosslesions are similar to those seen in other domestic species. In subclinical cases, a congested, occasionally nodular liver may be the only gross lesion as a result of chronic passive congestion in this organ. The heart may appear enlarged, and the right ventricle may appear thin or flabby. With progressively severe cases, there is often an accumulation of a serosanguinous ascitic transudate in the abdominal cavity, the pleural cavity, or both. In severe cases, the lungs are atelectatic and compressed by the presence of a globose heart and abundant pleural effusion. In cases in which the heart is not enlarged. The examination of the left ventricular free wall and the interventricular septum may reveal marked thickening and impingement upon the ventricular lumen. Rarely, the presence of fibrous connective tissue may be seen upon close inspection of the cardiac wall, and occasionally, due a previous ischemic event, a focally extensive area of the ventricular wall may be translucent and paper thin as a result of total loss of myocytes in this area and replacement by fibrous connective tissue.

Early lesions consist of an increase in fibrous connective tissue around myocardial vessels which extends into the interstitium. As the condition progresses, there is atrophy and loss of myocytes. Focal areas of myocyte degeneration may be present, with an infiltrate of moderate numbers of macrophages, lymphocytes, plasma cells, and rare neutrophils. In some cases of cardiomyopathy, there may be marked focal malalignment of myocytes, suggesting orientation in several different planes.

Centrilobular fibrosis, edema, micronodular hemosiderosis, and loss of subcapsular hepatocytes with resulting fibrosis all attest to chronic hepatic congestion, which is a common finding in cardiac disease in the ferret: In contrast, the presence of chronic signs of left-sided heart failure is relatively uncommon. In terminal stages of the disease, there may be necrosis of centrilobular hepatocytes due to stasis and hypoxia. The presence of marked myocardial fibrosis with or without inflammation, and evidence of chronic systemic congestion are highly suggestive of cardiomyopathy in this species.

2. Notes to the Text B

(1) Several forms of this condition may be seen-dilatative, hypertrophic, and a restrictive form in which there is marked replacement of myocardium by fibrous connective tissue, with minimal change in chamber area.

句中“in which”引导的定语从句，通常等于“where”，这里“in”不能省略。

（2）Rarely，the presence of fibrous connective tissue may be seen upon close inspection of the cardiac wall，and occasionally，due a previous ischemic event，a focally extensive area of the ventricular wall may be translucent and paper thin as a result of total loss of myocytes in this area and replacement by fibrous connective tissue.

句中“Rarely”和“occasionally”为否定副词，置于句首时，后面的句子要用部分倒装形式，这种一般用在正式文体中，这一类含有否定意义的副词有“never，seldom，little，hardly，scarcely，no sooner，no longer，nowhere”等。

（3）In some cases of cardiomyopathy，there may be marked focal malalignment of myocytes，suggesting orientation in several different planes.

句中“Suggesting”是“动词-ing”做宾语，“动词-ing”形式可以用作动词、短语动词和介词的宾语。

3. Answer Questions

（1）Is cardiomyopathy a common disease in the American lines of ferrets?

（2）What may be the only gross lesion in subclinical cases?

（3）What may be happpened in the severe cases of cardiomyopathy?

（4）What do early lesions consist of ?

（5）When did it happen there atrophy and loss of myocytes?

（6）Why may be necrosis of centrilobular hepatocytes in terminal stages of the disease?

课文B　雪貂心肌病

心肌病是美国雪貂常见的一种疾病，据推测有遗传基础。这种情况有几种表现形式——扩张性的、肥厚性的和限制性的，这些都明显表现为纤维结缔组织替代心肌，心室面积变化极小。如果动物1岁时严重感染，就会出现心肌病症状，但常见的是在5~7岁。

大体的病灶与其他家养物种相似。在亚临床病例中，充血、偶见因器官慢性被动充血导致的结节状的肝脏，可能是唯一的严重病变。心脏可能出现增大，右心室可能变薄或松弛。在逐渐严重的病例中，常常会在腹腔、胸膜腔或两者中积累浆液性腹水。在严重的病例中，肺因球形心脏和大量胸腔积液而通气不畅和扩张不全。在心脏没有扩大的情况下。对左心室游离壁和室间隔的检查可发现室腔壁明显增厚并影响心室腔。少量病例在近距离检查心室壁可能会看到纤维结缔组织的存在，偶尔由于前一个缺血性事件，局部心室壁的区域可能出现成片半透明，并且在这一区域有纸一样薄的全损的细胞，同时出现纤维化心肌细胞。

早期病变包括心肌血管周围的纤维结缔组织增加，并向间质延伸。随着病情的发展，肌细胞萎缩和丧失。可能存在局部的肌细胞变性，同时出现中度数量的巨噬细胞、淋巴细胞、浆细胞和罕见的中性粒细胞浸润。在某些心肌病病例中，可能会出现明显的心肌细胞病灶排列不整齐，提示在几个不同的平面上有定位。

小叶中心纤维化、水肿、微结节性含铁血黄素沉积、被膜下肝细胞减少导致的纤维化，证实出现慢性肝充血，这是貂心脏病中常见的症状：相对而言，左侧心衰的慢性体征相对少见。在疾病的终末期，可能出现小叶中心肝细胞坏死，这是由于停滞和缺氧所致。明显的心肌纤维化伴有或不伴有炎症，以及慢性全身充血的症状在雪貂中提示着有心肌病的发生。

Lesson 24

Zoonoses

Part One　Intensive Reading

1. Learning Objective

After learning this lesson, you should understand the following:

(1) Explain why Zoonoses are generally defined as animal diseases that are transmissible to humans.

(2) Explain why the emergence and reemergence of zoonotic diseases present challenges not only to veterinarians, but also to all professions concerned with public health.

(3) Describe people with acquired immune deficiency syndrome (AIDS) are much more susceptible to which in general.

(4) Explain why the plague bacterium may be more adept at finding new hosts or new foci.

(5) Give an overview of why knowledge of the epidemiology of the disease organisms is the first step in initiating a control program.

2. Text A　Zoonoses

Zoonoses are generally defined as animal diseases that are transmissible to humans. However, there are several diseases listed below that occur primarily in humans and that may also be transmitted between humans and animals, with some animals serving as reservoirs for human infection. The following common bacterial and viral diseases of humans are not found as naturally occurring diseases in animals (*i.e.* animals are not a reservoir): diphtheria, syphilis, typhoid fever, poliomyelitis, hepatitis B, smallpox, and measles.

The emergence and reemergence of zoonotic diseases present challenges not only to veterinarians, but also to all professions concerned with public health. Since the

19th century, the veterinary profession in the USA has been at the forefront of control and eradication of animal diseases, including bovine pleuropneumonia, foot-and-mouth disease, bovine tuberculosis, brucellosis, and classical swine fever. The early cooperation of veterinarians and public health physicians gave impetus to the eradication of bovine tuberculosis first in Denmark, Sweden, Finland, and Norway and then in the USA and Canada. Unfortunately, bovine tuberculosis has emerged in Mexico along the border with the USA and has caused human disease and dairy cow infections from California to Texas. Bovine tuberculosis and brucellosis remain major problems in the developing world.

People with acquired immune deficiency syndrome (AIDS) are much more susceptible, in general, to zoonotic diseases, including *tuberculosis*, *toxoplasmosis*, *cryptosporidial enteritis*, *Campylobacter*, and *Listeria*. It is possible that other zoonotic diseases that are dormant or infrequent (*e. g.*, *leptospirosis*, *plague*, and *glanders*) may emerge in individuals with AIDS or other immunocompromising conditions. Many of these are latent or nonpathogenic serovars. AIDS-like infections have been described in lions and the tropical cats of Africa as well as in domesticated cats. None have caused human disease.

In Australia and Malaysia, new diseases have been reported in horses and swine that also affect humans. They are caused by a morbillivirus-a measles-like virus related to canine distemper and rinderpest viruses. Another virus killed many wild felids in a zoo in Egypt. Many emerging viral diseases that have a rodent or unknown animal host have caused devastating fatal diseases in humans in Africa and South America, *e. g.*, Lassa fever, an arenavirus serologically related to lymphocytic choriomeningitis and South American hemorrhagic diseases of Argentina and Bolivia. In Africa, Ebola fever and Marburg disease, the latter a dormant monkey disease, have caused death in medical personnel and in patients. Crimean-Congo hemorrhagic fever has caused death in African travelers and in the Middle East in abattoir workers.

The death of veterinarians in the western USA from plague, and reports of serious illness in veterinary technicians and cat owners, has focused attention on both domestic and feral cats and the larger mountain lion or bobcats as carriers of this ancient disease. Dogs and wild canids are likewise involved in plague regions of the USA. The involvement of cats since the 1970s is evidence of the dynamics of zoonotic diseases in a changing environment. Human populations may be pressuring old habitats, or there may be more subtle changes. The plague bacterium may be more adept at finding new hosts or new foci, as seen in other emerging diseases.

The Hantaan virus complex was first noted in 1951 in Korea, where it caused a hemorrhagic disease with renal syndrome. Various forms of the disease exist worldwide. In the USA, two fatal forms of the disease have been reported, namely, nephritic and pneumonic (in addition to latent infections). Hanta virus has caused infections in laboratory rodents, and veterinary technicians have been infected in Asia and Europe.

Prion diseases have been described in North America, Europe, and Asia and are known to affect sheep, cattle, elk, and deer as well as wild and domestic felids. The bovine prion disease is reported to have caused more than 140 human cases with 100% mortality in the UK. Although the incidence is less than 1/1000000, the threat of prion diseases to human and animal health is of major concern.

Exposure to animals kept as pets is steadily increasing as the number of pets increases in the USA and other affluent countries. The types of animals kept as pets are also increasing. Examples of these are the "exotic pets" that have become popular in many parts of the world, *e.g.*, prairie dogs, that have brought plague, tularemia, and even monkeypox out of the wild into people's homes. The desire of humans to touch wild animals or have contact with farm animals has resulted in the establishment of "petting zoos." Contact with farm or wild animals may expose children or other visitors to organisms such as *Escherichia coli* O157: H4 or even rabies. Public health officials in the USA, Canada, and Britain are trying to control these "zoos" through inspections and rules, including microbicidal handwashing following exposures.

Another source of infection is exemplified by the severe acute respiratory syndrome epidemic caused by a novel coronavirus that appeared in southern China in 2003, first among food preparation workers exposed to civet cats and other "exotic animals" during their preparation as special foods.

The 21st century holds the threat of even more emerging diseases, nurtured by an everincreasing human population. Control of zoonotic diseases and protection of the public health will become even more challenging as world population increases. When overpopulation and crowding occur, water shortages occur, hygiene often cannot be maintained, and malnutrition develops, leading to disease and epidemics. Surveillance and reporting of disease is the first line of defense.

Knowledge of the epidemiology of the disease organisms is the first step in initiating a control program. The ultimate objective is to protect and preserve both human and animals health.

3. New Words and Phrases

zoonosis [ˌzuːəˈnəʊsɪs] n. 人畜共患病
pleuropneumonia [ˌplʊərəʊnjuːˈməʊnɪə] n. 胸膜性肺炎
foot-and-mouth disease 口蹄疫
bovine tuberculosis [ˈbəʊvaɪntjʊˌbɜːkjʊˈləʊsɪs] n. 牛结核病
brucellosis [ˌbruːsəˈləʊsɪs] n. 布鲁菌病
classical swine fever 古典猪瘟
acquired immune deficiency syndrome（AIDS）获得性免疫缺陷综合征
toxoplasmosis [ˌtɒksəʊplæzˈməʊsɪs] n. 弓形虫病
cryptosporidial enteritis 隐孢子虫肠炎
leptospirosis [ˌleptəʊspaɪˈrəʊsɪs] n. 钩端螺旋体病
plague [pleɪg] n. 鼠疫
glanders [ˈglændəz] n. 鼻疽
reservoir [ˈrezəvwɑːr] n. 水库，蓄水池
immunocompromising 免疫损害
serovar 血清型
latent [ˈleɪtənt] adj. 潜伏的
rinderpest [ˈrɪndəpest] n. 牛瘟
morbilli virus 麻疹病毒
canine distemper 犬瘟热
felid [ˈfiːlɪd] n. 猫科动物
choriomeningitis [ˌkəuriəuˌmeninˈdʒaitis] n. 脉络丛脑膜炎
arenavirus [əˈriːnəˌvaɪrəs] n. 沙粒病毒
hemorrhagic fever 出血热
feral [ˈferəl] adj. 野生的
tularemia [ˌtʊləˈrimɪə] n. 兔热病
coronavirus [kəˌrəunəˈvaiərəs] n. 冠状病毒
civet cat 果子狸
nurture [ˈnɜːtʃə] n. 养育，培养
diphtheria [dɪfˈθɪərɪə] n. 白喉
syphilis [ˈsɪfɪlɪs] n. 梅毒
typhoid [ˈtaɪfɒɪd] n. 伤寒
hepatitis B B型肝炎

measles [ˈmiːzəlz] *n.* 麻疹
alcoholism [ˈælkəhɑːlɪzəm] *n.* 酒精中毒
focus [ˈfəʊkəs] *n.* 疫源地
bobcat [ˈbɒbkæt] *n.* 美洲野猫
canid [ˈkænid] *n.* 犬科动物
louping-ill virus 跳跃病病毒
Hantaan virus 汉坦病毒

4. Notes to the Text A

(1) However, there are several diseases listed below that Occur primarily in humans and that may also be transmitted between humans and animals, with some animals serving as reservoirs for human infection.

这是一个复合句，主句是“there are several diseases”，句中有两个“that”引导定语从句，修饰“diseases”；“with some animals…”做状语进一步补充前面的句子。

译文如下：然而，下面文中列举的一些疾病首先发生在人身上，可以在人和动物之间传播，由某些动物作为人传染病的中间宿主。

(2) People with acquired immune deficiency syndrome (AIDS) are much more susceptible, in general, to zoonotic diseases, including *tuberculosis*, *toxoplasmosis*, *cryptosporidial enteritis*, *Campylobacter*, and *Listeria*.

此句是一个简单句，“in general”是插入语，“including”后的成分进一步解释说明“diseases”。

译文如下：患有获得性免疫缺陷综合征（艾滋病）的人通常对人畜共患病（如结核病、弓形虫隐孢子虫性肠炎、弯曲杆菌和李氏杆菌等）更为易感。

(3) Another source of infection is exemplified by the severe acute respiratory syndrome epidemic caused by a novel coronavirus that appeared in southern China in 2003, first among food preparation workers exposed to civet cats and other “exotic animals” during their preparation as special foods.

这是一个复杂的简单句，“Another source of infection”做主语，“caused”是过去分词，做后置定语，修饰“epidemic”；“that”引导定语从句，修饰“coronavirus”；“among”引导的介词短语起补充作用，其中“exposed to”引导定语成分，修饰“workers”。

译文如下：另一个传染源事例是，2003 年中国华南出现的一种新型冠状病毒所引起的、急性严重呼吸综合征的流行，首先在接触果子狸和其他“外来动物”（作为特殊食品）的食品加工工人中传播。

5. Exercises

(1) Fill in the blanks by finishing the sentences according to the passage

①Overpopulation and crowding will result in ________ .

②Since the 19th century, the veterinary profession in the USA has been at the forefront of control and eradication of animal diseases, including ________ .

③The early cooperation of veterinarians and public health physicians gave impetus to the eradication of bovine tuberculosis first in ________ .

④People with acquired immune deficiency syndrome (AIDS) are much more susceptible, in general, to zoonotic diseases, but not including ________ .

⑤AIDS-like infections have been described in ________ .

⑥In Australia and Malaysia, new diseases have been reported in ________ .

⑦The severe acute respiratory syndrome epidemic caused by a novel coronavirus appeared in ________ in 2003. and was found first among ________ . The Hantaan virus complex was first noted in ________ .

(2) Answer the following questions according to the passage

①How was bovine tuberculosis eradicated first in Denmark, Sweden, Finland, and Norway and then in the USA and Canada?

② Do bovine tuberculosis and brucellosis remain major problems in the developing world?

③Why do incidences of zoonotic diseases in human gradually increase?

④Why does the 21st century hold the threat of even more emerging diseases?

⑤What will be the results of overpopulation and crowding?

⑥What is the first line of disease defense?

⑦What is the first step in initiating a control program?

⑧What is the ultimate objective of controlling zoonoses?

⑨What do zoonoses refer to?

(3) Translation of the following sentences into Chinese

①Unfortunately, bovine tuberculosis has emerged in Mexico along the border with the USA and has caused human disease and dairy cow infections from California to Texas.

②AIDS-like infections have been described in lions and the tropical cats of Africa as well as in domesticated cats.

③Dogs and wild canids are likewise involved in plague regions of the USA.

④The Hantaan virus complex was first noted in 1951 in Korea, where it caused a hemorrhagic disease with renal syndrome.

⑤Public health officials in the USA, Canada, and Britain are trying to control these "zoos" through inspections and rules, including microbicidal handwashing following exposures.

⑥Another source of infection is exemplified by the severe acute respiratory syndrome epidemic caused by a novel coronavirus that appeared in southern China in 2003, first among food preparation workers exposed to civet cats and other "exotic animals" during their preparation as special foods.

⑦Control of zoonotic diseases and protection of the public health will become even more challenging as world population increases.

课文A 人畜共患病

人畜共患病通常是指传播给人的动物疾病。然而，下面文中列举的一些疾病首先发生在人身上，可以在人和动物之间传播，由某些动物作为人传染病的中间宿主。下列人类常见的细菌和病毒病并不在动物中自然发生（即动物不是这些疾病的中间宿主）：白喉、淋病、伤寒热、脊髓灰质炎、乙型肝炎、天花和麻疹等。

人畜共患病的出现和再现不仅给兽医工作者，而且给所有公众卫生相关行业带来了挑战。自19世纪以来，美国的兽医行业一直是控制和消除动物疾病（如牛胸膜性肺炎、口蹄疫、牛结核病、布鲁菌病以及古典猪瘟等）的前线。兽医师和公众卫生医师的早期合作促进了丹麦、瑞典、芬兰和挪威首先消灭牛结核，然后美国和加拿大也消灭此病。遗憾的是，沿着墨西哥与美国边界出现了牛结核，并又导致从加利福尼亚州到得克萨斯州的人和奶牛发病。牛结核和布鲁菌病仍是发展中国家面临的严重问题。

患有获得性免疫缺陷综合征（艾滋病）的人通常对人畜共患病（如结核病、弓形虫病、隐孢子虫性肠炎、弯曲杆菌和李氏杆菌等）更为易感。处在休眠状态或不常发生的人畜共患病（即钩端螺旋体病、瘟疫和鼻疽等）很可能在患艾滋病或其他免疫不全的病人中出现。这些病中许多呈潜伏态或是非致病血清型。在非洲狮和热带猫以及家猫中发现了艾滋病样疾病，但尚未使人发病。

据报道，在澳大利亚和马来西亚，马和猪出现一些新型疾病，也能影响人。它们是由一种麻疹病毒引起，这种麻疹样病毒与犬瘟热和牛瘟病毒的相关。在埃及一家动物园，另一种病毒导致了许多野生猫科动物的死亡。有许多正在出现的病毒，其宿主是啮齿类或不明动物，它们在非洲和南美洲引起了毁灭性的疾病，如拉萨热，它由一种沙粒病毒引起，该病毒的血清型与淋巴细胞性脉络丛脑膜炎病毒以及阿根廷和玻利维亚的南美洲出血病病毒相关。非洲的埃博拉热和马尔堡病已引起医务工作者和病人死亡，后一种是潜伏的猴病。克里米亚，刚果出血热引起了在非洲的旅游者和中东屠宰场工人死亡。

美国西部兽医死于瘟疫，以及兽医技师和猫主人患严重疾病的报道，使我们把注意力聚焦在家猫、野猫以及大型山狮或山猫上，它们是这种古老疾病的携带者。犬和野生犬科动物同样与美国瘟疫的传播有关。自 20 世纪 70 年代证明猫参与人畜共患病以来，人畜共患病可以传播到其他地方。人口增多对动物栖息地造成的压力（而移居他处）可能是其原因，此外或许还存在一些更加微妙的变化。瘟疫病原，类似于其他已经出现的疾病那样，可能会更擅长发现新的宿主或新的疫源地。

1951 年首次发现汉坦病毒复合体，它引起了带有肾脏综合征的出血病。全球存在不同种类的汉坦病毒病。据报道，美国有两种致命型，即肾型和肺型（除了潜伏传染之外）。汉坦病毒已感染一些啮齿类实验动物，并且亚洲、欧洲的兽医技师也有被感染的。

在北美、欧洲和亚洲均出现了朊病，已知它可影响绵羊、牛、麋和鹿，以及家养和野生猫科动物。据报道，牛朊病已引起英国 140 多人发病，死亡率 100%。虽然发生率不到百万分之一，但朊病对人和动物健康的威胁已是人们很关心的事。

随着美国和其他富裕国家宠物数量的增加，与宠物接触的机会也随之增加。作为宠物的动物类型也在增加。“外来宠物”在世界许多地区变得流行，如草原土拨鼠，这就把野外的鼠疫、兔热病甚至猴痘等带到家中。人们接触野生动物或农畜的愿望导致“宠物园”的建立。与农畜或野生动物的接触可能使孩子或游客接触到 O157：H4 埃希大肠杆菌，甚至狂犬病病毒等。美国、加拿大和英国的公共卫生官员试图通过检查和制定条例的方法来控制“宠物园”，其中包括接触动物后要用杀菌剂洗手。

另一个传染源事例，2003 年中国华南出现的一种新型冠状病毒所引起的、急性严重呼吸综合征的流行，它首先在接触果子狸和其他“外来动物”（作为特殊食品）的食品加工工人中传播。

随着人口数量的持续增加，21 世纪人类面临着更多新型疾病的威胁。随着全球人口的增长，人畜共患病的控制和公共卫生的保护将面临更大的挑战。人口过剩和拥挤将导致缺水、卫生得不到保护和营养缺乏，这最终导致疾病的发生和流行。监督和报道疾病是第一道防御线。了解病原微生物的流行病学是启动控制计划的第一步。最终目的就是保护和维持人类和动物健康。

Part Two Extensive Reading

1. Text B Prions and Transmissible Spongiform Encephalopathies

Prions are not virus and appear to be a new class of infectious agent that lead to chronic progressive infections of the nervous system, inducing common pathological effects, the results of which, after a long incubation period of perhaps several years, are invariably fatal. The clinical syndromes for which they are responsible are known collectively as transmissible spongiform encephalopathies (TSEs). Examples, in animals, include scrapie in sheep, transmissible mink encephalopathies (TME), feline spongiform encephalopathy (FSE) and bovine spongiform encephalopathy (BSE or "mad COW" disease). There are four forms of human TSEs: Creutzfeldt Jakob disease (CJD), fatal familial insomnia (FFI), Gerstmann - Strausster - Scheinker disease (GSS) and Kuru.

There has been much speculation as to the molecular nature of the infectious agents responsible for these disease conditions, but in 1972 Stanley Prusiner penned the term prion (proteinaceous infectious particle PrP). His hypothesis was that these agents are totally free of nucleic acid and consist solely of protein, a hypothesis for which much supportive evidence has accumulated. In 1997 Prusiner was awarded the Nobel Prize for his studies.

Whilst the various conditions induced by prions have subtle differences there are many features in common. All have long incubation periods, with the agent replicating in a number of tissues, *e.g.*, spleen, liver, before appearing in high concentrations in the brain and CNS towards the terminal stages of the disease. The agents are responsible for severe degeneration of the brain and spinal cord and, once clinical signs appear, the condition is invariably fatal. Pathologically the disease is characterized by the appearance of abnormal protein deposits (amyloids), in various tissues, *e.g.*, kidney, spleen, 1iver and brain. Amyloids arise from the accumulation of various proteins, which take the form of plaques or fibrils.

Amyloidosis is also a characteristic feature of *e. g.* , Alzheimer's disease, although this condition is not transmissible. The protein deposits are cytotoxic and are responsible for the resultant pathology, namely the sponge-like appearance of the brain where small holes can be visualized by microscopy in thin sections of brain tissue taken at post-mortem. It is this appearance which gave the spongiform encephalopathies their name. There is no conventional immune response to the agent, although the immune system plays an important part in the development of the disease before the agent gets into the CNS.

A diagnosis of TSEs can only be made by demonstrating the presence of prion proteins in tissue taken at post-mortem. This is usually by immunohistochemical stainingof the prion protein or by transfer of the agent to permissive experimental animals.

Animal TSEs Scrapie has been recognized as a distinct infection in sheep for over 250 years. A major investigation into its etiology followed the vaccination of sheep for louping-ill virus with formalin-treated extracts of ovine lymphoid tissue, unknowingly contaminated with scrapie prions. Two years later, more than 1500 sheep developed scrapie from this vaccine. The scrapie agent has been extensively studied and experimentally transmitted to a range of laboratory animals, *e. g.* , mice and hamsters. Infected sheep show severe and progressive neurological symptoms, such as abnormal gait. The name owes itself to the fact that sheep with the disorder repeatedly scrape themselves against fences and posts. The natural mode of transmission between sheep is unclear although it is readily communicable in the flocks. The placenta has been implicated as a source of prions, which could account for horizontal spread within flocks. In Iceland scrapie-infected flocks of sheep were destroyed and the pastures left vacant for several years. However, reintroduction of sheep from flocks known to be free of scrapie for many years eventually resulted in scrapie! Sheep have also been infected by feed-stuff contaminated with BSE, to which they are susceptible.

Bovine spongiform encephalopathy (BSE) is also called "mad cow" disease which appeared in Great Britain in 1986 as a previously unknown disease. Affected cattle showed altered behavior and a staggering gait. Post-mortem revealed protease-resistant PrP in the brains of the cattle and the typical spongiform pathology. To date >190000 cattle have been infected with this agent which, it is thought, initiated from the common source of contaminated meat and bone meal (MBM) given to cattle as a nutritional supplement. MBM was initially prepared by rendering the offal of sheep and cattle using a process that involved steam treatment and hydrocarbon solvent extraction. In the late 1970s, however, the solvent procedure was eliminated from the process, resulting in high concentrations of fat in the MBM; it is postulated that this high fat content protected scrapie prions in the sheep offal from being completely inactivated by the steam. Thus the initial MBM contained only sheep prions, and the similarity between bovine and sheep PrP was probably an important factor in initiating the BSE epidemic. Bovine PrP differs from sheep PrP at 7 or 8 residues. As the BSE epidemic expanded, infected bovine offal began to be rendered into MBM that contained bovine prions, this being fed to cattle!

Since 1988 the practice of using dietary protein supplements for domestic animals derived from rendered sheep or cattle offal has been forbidden in the UK. Statistics argue that this food ban has been effective in getting the epidemic under control.

Brain extracts from BSE cattle have transmitted disease to mice, cattle, sheep and pigs after intracerebral inoculation. Disease has also followed in mink, domestic cats, pumas and cheetahs after oral consumption of BSE prions in food-stuffs.

Evidence is accumulating that the BSE prion has been transferred to humans and is responsible for variant CJD (vCJD). The oral route is suspected. The source of the infected material is unknown, although BSE-contaminated cattle products are the probable source. In 1989 the human consumption of specified bovine offals (brain, spleen, thymus, tonsil, gut) was prohibited in the UK. In 1996 this ban was extended to sheep. Furthermore, the brain and CNS are now removed from all cattle at abattoirs prior to the distribution of meat.

2. Notes to the Text B

(1) Prions are not virus and appear to be a new class of infectious agent that lead to chronic progressive infections of the nervous system, inducing common pathological effects, the results of which, after a long incubation period of perhaps several years, are invariably fatal.

这是一个复合句。“Prions”做主语，有两个谓语，分别是“are not virus”和“appear to be a new class of infectious agent”；“that”引导定语从句，修饰“a new class of infectious agent”；“which”引导定语从句，指“common pathological effects”；“after a long…”做插入语。

译文如下：朊不是病毒，它似乎是一种新型的有传染性的物质，能够导致神经系统慢性、渐进性感染，引起常见的病理效应，因此经过长时间，或许几年的潜伏期后，最终导致死亡。

（2） His hypothesis was that these agents are totally free of nucleic acid and consist solely of protein, a hypothesis for which much supportive evidence has accumulated.

这是一个复合句，“His hypothesis was that”，“hypothesis”做主语，“that”引导的表语从句，后面的“a hypothesis”是一个同位语，“which”引导定语从句，修饰“a hypothesis”。

译文如下：（斯坦利·普鲁森）他假设这些物质全无核酸，仅由蛋白质组成。支持这种假设的证据越来越多。

（3） A major investigation into its etiology followed the vaccination of sheep for louping-ill virus with formalin-treated extracts of ovine lymphoid tissue, unknowingly contaminated with scrapie prions.

这是一个简单句。“investigation”做主语；“followed”为谓语动词，“the vaccination of sheep”做宾语，这个宾语比较复杂，“for louping-ill virus”表示为什么病毒接种疫苗，“formalin-treated extracts of ovine lymphoid tissue”表示接种什么样的疫苗，“contaminated with scrapie prions”修饰“extracts of ovine lymphoid tissue”。

译文如下：一项对痒病病因学的重点研究出现在给绵羊接种预防羊脑脊髓炎病毒的疫苗之后。该疫苗使用了福尔马林浸泡过的绵羊淋巴组织提取液，却不知该提取液已被痒病朊所污染。

3. Answer Questions

（1） Why did Stanley Prusiner pen the term prion?

（2） For what did Stanley Prusiner receive the Nobel Prize in 1997?

（3） Please explain how did the name spongiform encephalopathies come into being.

（4） What is the origin of the name for the disease called sheep scrapie?

（5） Why didn't MBM prepared prior to the late 1970s cause scrapie or mad cow disease?

（6） How does UK now control and prevent TSE of cattle and sheep?

课文B 朊和传染性海绵状脑炎

朊不是病毒，它似乎是一种新型的有传染性的物质，能够导致神经系统慢性、渐进性感染，引起常见的病理效应，因此经过长时间，或许几年的潜伏期后，最终导致死亡。此病的临床综合症状是传染性海绵状脑病（TSEs）。动物的TSEs包括绵羊痒病、貂传染性脑病（TME）、猫海绵状脑病（FSE）和牛海绵状脑病（BSE或“疯牛病”）。人的TSEs有四种：克罗伊茨费尔特-雅各布病（CJD）、致命性家族失眠症（FFI）、格斯特曼综合征（GSS）和新几内亚震颤病。

关于引起这些疾病的传染因子的分子特性有许多推测，但在1972年斯坦利·普鲁森用了朊一词（蛋白质样传染颗粒，PrP）。他假设这些物质全无核酸，仅由蛋白质组成。支持这种假设的证据越来越多。1997年普鲁森因此研究获得诺贝尔奖。

朊所致疾病的症状虽有微小差别，但更有许多共同特点。它们都有较长的潜伏期，当病到末期，朊在脑部和中枢神经系统（CNS）大量出现之前，可在许多组织（如脾和肝）中复制。朊可使脑和脊髓严重变性，一旦出现临床症状，结果就是死亡。此病的病理学特点是在许多组织，如肾、脾、肝和脑中，出现蛋白质异常沉积（淀粉体）。淀粉体是各种蛋白质积聚的结果，呈空斑或原纤维状。淀粉样变性病也是阿尔茨海默病的典型特点，但阿尔茨海默病并不传染。蛋白质沉积具有细胞毒性，可引起病理变化，即脑部出现海绵样变，尸体剖检取脑组织作切片用显微镜检查可以发现小洞。海绵状脑病由此得名。尽管在致病物质进入CNS以前，免疫系统在此病的发生上起重要作用，但机体对此物质没有常规的免疫应答反应。

TSEs的诊断只能靠检验尸检获取的组织样中是否有朊蛋白。方法通常包括朊蛋白的免疫组化染色或将其转染给易感实验动物。

动物TSEs中，痒病被认为是绵羊的独特疾病已有250多年了。一项对痒病病因学的重点研究出现在给绵羊接种预防羊脑脊髓炎病毒的疫苗之后。该疫苗使用了福尔马林浸泡过的绵羊淋巴组织提取液，却不知该提取液已被痒病朊所污染。两年以后，1500多只绵羊因接种此疫苗而患痒病。目前已对痒病病因物质进行了广泛研究，并试验性地转染给实验动物，如小鼠和仓鼠等。感染的绵羊表现严重的渐进性的神经型症状，如步履蹒跚。痒病名字的来源还归因于以下事实，即患此病的绵羊经常在栅栏和柱子上摩擦。虽然此因子很容易在羊群中传播，但绵羊之间的自然传播方式尚不清楚。胎盘也是朊的传播源，这能解释该因子在羊群内的水平传播。在冰岛，患痒病的绵羊群被消灭，牧场停止放牧几年。然而，从被认为多年未患痒病的羊群中引进的绵羊最终还是发生了痒病！绵羊还可被BSE污染的饲料所感染，绵羊对经此传播的BSE易感。

牛海绵状脑病（BSE）又称“疯牛病”，1986年在英国出现，这是一种以前尚不知道的疾病。患病牛行为发生改变，步履蹒跚。尸体剖检可见，牛脑部出现抗蛋白质酶的PrP和典型的海绵状病变。迄今为止，共有190000多头牛被此致病因子感染。一般认为，这些牛是因吃了作为营养添加剂的、普通来源的、被污染的肉骨粉（MBM）而发病的。MBM最初的制备方法是将屠宰牛和羊的废弃料用蒸汽处理和烃类溶剂提取。然而，在20世纪70年代末期，MBM的加工过程中取消了溶剂提取步骤，致使MBM中脂肪浓度过高；可以推论，高浓度脂肪保护了羊屠宰碎料中的痒病朊免受蒸汽的彻底灭活。因此，最初MBM中仅含有绵羊朊，而牛和绵羊PrP的类似性很可能是导致BSE流行的重要因素。牛PrP在第7或第8残基处不同于绵羊PrP。当BSE流行范围扩大时，感染牛的下水被制成MBM，而MBM含有牛朊并用来喂牛。

自1988年以来，英国已禁止使用绵羊和牛下水作为家畜饲料蛋白质添加剂。数据证明该饲料禁令对控制BSE的流行是有效的。

脑内接种BSE牛脑组织提取物可以把疾病传播给小鼠、牛、绵羊和猪。貂、家猫、美洲豹和印度豹在进食含BSE朊的饲料后也发病了。

BSE朊可以传播给人并引起变异克罗伊茨费尔特-雅各布病的证据在增加。人们猜测该病是经口传播的。尽管BSE感染的牛产品可能是疾病的来源，但感染的材料来源尚不清楚。1989年英国禁止人食用特定的牛下水（脑、脾、胸腺、扁桃体和肠）。1996年此禁令扩展到绵羊下水。而且，屠宰场在肉出库以前就去除所有牛的脑和中枢神经系统。

Appendix 附录

Vocabulary
词汇表

A

abdominal [æbˈdɑːmɪnl] *adj.* 腹部的
abiotic [æbˈdɑːmɪnl] *adj.* 无生命的，非生物的
abnormality [ˌæbnɔːrˈmæləti] *n.* 异常
abscess [ˈæbses] *n.* 脓肿
absorb [əbˈzɔːrb，əbˈsɔːrb] *vt.* 吸收；吸引；承受；理解；使……全神贯注
absorbability [əbˌsɔrbəˈbɪləti] *n.* 吸收，可吸收性；被吸收
absorption [əbˈzɔːrpʃn，əbˈsɔːrpʃn] *n.* 吸收；全神贯注，专心致志
accommodation [əˌkɑːməˈdeɪʃn] *n.* 住处；适应；和解；便利
accuracy [ˈækjərəsi] *n.* 精确（性），准确（性）
acetic acid 醋酸
acetoacetic acid 乙酰乙酸
acetone body 酮体
acid-base equilibrium 酸碱平衡
acquired immune deficiency syndrome（AIDS）获得性免疫缺陷综合征
acquired immunity 获得性免疫
acreage [ˈeɪkərɪdʒ] *n.* 面积，英亩数
actin [ˈæktɪn] *n.* 肌动蛋白，肌纤蛋白
activation [ˌæktɪˈveɪʃn] *n.* 激活；活化作用

adaptive immunity 适应性免疫

adenylic acid 腺苷酸

adequate [ˈædɪkwət] *adj.* 足够的；适当的，恰当的；差强人意的；胜任的

adipose tissue 脂肪组织

air-dry feed 风干饲料

alcohol [ˈælkəhɔːl] *n.* 酒精，乙醇

alcoholism [ˈælkəhɑːlɪzəm] *n.* 酒精中毒

alfalfa [ælˈfælfə] *n.* 苜蓿；紫花苜蓿

alkali [ˈælkəlaɪ] *n.* 碱；可溶性无机盐；*adj.* 碱性的

allergy [ˈælərdʒi] *n.* 过敏

alleviate [əˈliːvieɪt] *vt.* 减轻，缓和

along with 连同；以及；和……一起

alter [ˈɔːltər] *vt.* 改变，更改；*vi.* 改变；修改

alternative [ɔːlˈtɜːrnətɪv] *adj.* 供选择的；选择性的；交替的；*n.* 二择一；供替代的选择 .

amine [əˈmin] *n.* 胺

ammonia [əˈmoʊniə] *n.* 氨

anaerobic [ˌænəˈroʊbɪk] *adj.* 厌氧的，厌气的；没有气而能生活的

anatomy [əˈnætəmi] *n.* 解剖；解剖学；剖析；骨骼

anemia [əˈnimiə] *n.* 贫血；贫血症

antibody [ˈæntibɑːdi] *n.* 抗体

antigen clearance 抗原清除

antigen presentation 抗原呈递

antiscurvy [ˌæntiˈskɜːrvi] 抗坏血病

APCs antigen-presenting cells 抗原呈递细胞

apparent [əˈpærənt] *adj.* 易看见的，可看见的；显然的，明明白白的；貌似的，表面的；显见

appetite [ˈæpɪtaɪt] *n.* 欲望；胃口，食欲；嗜好，爱好

arenavirus [əˈriːnəˌvaɪrəs] *n.* 沙粒病毒

arginine [ˈɑrdʒəˌnin] *n.* 精氨酸

artery [ɑːrtəri] *n.* 动脉，干道，主流

arthritis [ɑːrˈθraɪtɪs] *n.* 关节炎

artificially inseminate 人工授精

ascite [æˈsaɪt] 腹水

ash [æʃ] *n.* 粗灰分；灰；灰烬

assistance [əˈsɪstəns] *n.* 帮助，援助

assumed [əˈsuːmd] *v.* 假定；假设；取得（权力）assume 的过去式和过去分词；呈现

at first glance 乍一看，看一眼

autochthonous [ɔːˈtɑːkθənəs] *adj.* 当地的；土著的；独立的；地方性的

autoimmune [ˌɔːtoʊɪˈmjuːn] *adj.* 自体免疫的

availability [əˌveɪləˈbɪləti] *n.* 可用性；有效性；实用性

average daily gain 平均日增重

avian [ˈeɪviən] *adj.* 鸟的，鸟类的

axon [ˈæksɒn] *n.* 轴突

B

barley [ˈbɑːrli] *n.* 大麦

battery [ˈbætəri] *n.* 电池；一系列；炮兵连；排炮；[律] 殴打

be associated with 与……有关

beasts of burden 役用动物

bee venom 蜂毒

beneficial [ˌbenɪˈfɪʃl] *adj.* 有利的，有益的

benzene [ˈbenziːn] *n.* 苯

bermudagrass [bəˌmjʊdəˈgræs] 狗牙根草

biomedical [ˌbaɪoʊˈmedɪkl] *adj.* 生物医学的

biotic [baɪˈɑːtɪk] *adj.* 有关生命的；生物的

bird's-foot trefoil 百脉根

bladder [blædər] *n.* 膀胱

blastocyst [ˈblæstəsɪst] *n.* 胚泡，囊胚

bleach [bliːtʃ] *n.* 漂白，漂白剂

blood clotting 血液凝固

bloodstream [ˈblʌdstriːm] *n.* 血流，血液的流动

blow-fly 绿头苍蝇

bluegrass [ˈbluːgræs] *n.* 莓系属的牧草；早熟禾属植物

bobcat [ˈbɒbkæt] *n.* 美洲野猫

body temperature 体温

body weight 体重

bone [bəʊn] *n.* 骨；骨骼；*vt.* 剔去……的骨；施骨肥于；*vi.* 苦学；专心致志

bovine tuberculosis [ˈbəʊvaɪn] [tjʊˌbɜːkjʊˈləʊsɪs] *n.* 牛结核病

branch [bræntʃ] *n.* 分部；部门；分店；分支；树枝 *v.* 分岔；分支

breakthrough [brekˈθrʊ] *n.* 发现，突破

breed [briːd] *vi.* 繁殖；饲养；产生；*vt.* 繁殖；饲养；养育，教育；引起；*n.* [生物] 品种；种类，类型

breeding stock 种畜

brief [briːf] *adj.* 短暂的，简短的，简要的，简略的；*n.* 简报，短简；*v.* 交代

bromegrass [ˈbromgræs] *n.* 雀麦草；雀麦属植物

brucellosis [ˌbruːsəˈləʊsɪs] *n.* 布鲁菌病

buffalograss 野牛草

buildup [ˈbɪldʌp] *n.* 集结；增长；树立名誉

bummer [ˈbʌmər] *n.* 失败

butyric acid 丁酸

β-hydroxybutyric acid β-羟基丁酸

C

calcium [ˈkælsɪəm] *n.* 钙

calcium salt 钙盐

calf [kɑːf] *n.* [解剖] 腓肠，小腿；小牛；小牛皮；(鲸等大哺乳动物的) 幼崽

canaliculi [ˌkænəˈlikjuliː] *n.* 小管，细管，微管 canaliculus 的复数

canary grass 加那利草

canid [ˈkænid] *n.* 犬科动物

canine distemper 犬瘟热

cannibalism [ˈkænəblˌɪzəm] *n.* 吃人肉的习性；同类相食

carbohydrate [kɑːbəˈhaɪdreɪt] *n.* 碳水化合物；糖类

carbon dioxide 二氧化碳

carcass [ˈkɑrkəs] *n.* (人或动物的) 尸体；(家畜屠宰后的) 躯体；骨架，遗骸，残迹等

cardiac [ˈkɑːdɪæk] *adj.* 心脏的

cardiac muscle fiber 心肌纤维

cartilage [ˈkɑːrtɪlɪdʒ] *n.* 软骨

catalytic agent 催化剂

cater [ˈketər]，*vt.* & *vi.* 提供饮食及服务，*vt.* 满足需要，适合；投合，迎合

cavity [ˈkævɪtɪ] *n.* 腔；窝洞；蛀牙，龋洞

ceaselessly [ˈsiːslisli] *adv.* 不停地，持续地；不住

cecal [ˈsiːkl] *adj.* 盲肠的

cell body 细胞体

certified [ˈsɜːtəˌfaɪd] *adj.* 被鉴定的；被证明的；有保证的；公认的

cervix [ˈsɜːvɪks] *n.* 子宫颈；颈部

chemical [ˈkɛmɪkl] *n.* 化学制品，化学药品；*adj.* 化学的

chemotherapy [ˌkiːmoʊˈθerəpi] *n.* 化学疗法，化疗

chew [tʃuː] *vt.* & *vi.* 咀嚼，咬；深思，考虑 *n.* 咀嚼；咀嚼物

chlorine [ˈklɔːriːn] *n.* 氯

chloroform [ˈklɔːrəfɔːm] *n.* 氯仿；三氯甲烷；*vt.* 用氯仿麻醉

cholesterol [kəˈlestərɒl] *n.* 胆固醇

choriomeningitis [ˌkəuriəuˌmeninˈdʒaitis] *n.* 脉络丛脑膜炎

circulation [sɜːkjʊˈleɪʃən] *n.* 流通，传播；循环；发行量

circumstance [ˈsɜːrkəmstæns] *n.* 环境，境遇；事实，细节；典礼，仪式

citric acid 柠檬酸

civet [ˈsɪvɪt] cat 果子狸

classical swine fever 古典猪瘟

classify [ˈklæsɪfaɪ] *vt.* 分类；归类

clonal selection 克隆选择

clostridial [klɔsˈtridiəl] *adj.* 梭菌的，梭菌属的

clostridiumperfringens [klɑsˈtrɪdɪəmˈpɜːfrɪndʒenz] 产气荚膜梭菌

clotting [ˈklɔtiŋ] *n.* 凝血；结块；*v.* 结块（clot 的 ing 形式）

collagen [ˈkɒlədʒən] *n.* 胶原，胶原质

colon [ˈkəʊlən] *n.* 结肠

colonize [ˈkɒlənaɪz] *vt.* 将……开拓为殖民地；从他地非法把选民移入；移于殖民地；[生] 移植植物 *vi.* 开拓殖民地；移居于殖民地

colostrum [kəˈlɒstrəm] *n.* 初乳

columnar epithelium 柱形上皮

commercial [kəˈmɜːrʃl] *adj.* 商业的；贸易的；盈利的

comparison [kəmˈpærɪsən] *n.* 比较，对照

complaints [kəmˈpleɪnts] *n.* 投诉；抱怨

component [kəmˈpəʊnənt] *adj.* 组成的，构成的；*n.* 成分；组件；[电子] 元件

comprise [kəmˈpraɪz] *vt.* 包括；由……组成

conceive [kən'siːv] *vt.* 怀孕；构思；以为；持有；*vi.* 怀孕；设想；考虑

concentrates ['kɑnsɛnˌtret] *n.* 精矿；浓缩液（concentrate 的复数）；精料；浓缩物

conception rate 受孕率

concern [kən'sɜːn] *vt.* 涉及，关系到；使担心 *n.* 关系；关心；关心的事；忧虑

condemnation [ˌkɑːndem'neɪʃn] *n.* 谴责；定罪；谴责（或定罪）的理由；征用

confinement [kən'faɪnmənt] *n.* 关押；分娩；限制，约束

connective tissue 结缔组织

consciousness ['kɑnʃəsnəs] *n.* 意识；知觉；自觉；觉悟

consist of [kən'sɪst ʌv] 包括；由……组成

consultant [kən'sʌltənt] *n.*（受人咨询的）顾问；会诊医生

consumption [kən'sʌmpʃən] *n.* 消费；消耗；肺痨

contact ['kɑːntækt] *v.* 联系；接触

contemporary [kən'tempəreri] *adj.* 当代的；属一个时期的

continuously [kən'tɪnjʊəslɪ] *adv.* 连续不断地，接连地；时时刻刻；连着；直

conventional [kən'vɛnʃənəl] *adj.* 传统的；平常的；依照惯例的

convert [kən'vɜːrt] *v.* 转变；转换

coronavirus [kəˌrəʊnə'vaiərəs] *n.* 冠状病毒

county agent 农区指导员

creatine phosphate 磷酸肌酸

crested wheatgrass 冠毛大麦草

critter ['krɪtər] *n.* 动物

cryptosporidial enteritis 隐孢子虫肠炎

cuboidal epithelium 立方形上皮

curious ['kjʊəriəs] *adj.* 好奇的；奇妙的；好求知的；稀奇的

cystine ['sɪstiːn] *n.* 胱氨酸；双硫丙氨酸

cytokine ['saɪtəkaɪn] *n.* 细胞因子，细胞活素类物质

cytotoxic [saɪtə'tɑksɪk] *adj.* 细胞毒素的

D

dairy ['deərɪ] *n.* 乳制品；乳牛；制酪场；乳品店；牛乳及乳品业；*adj.* 乳品的；牛乳的；牛乳制的；产乳的

dandelion [ˈdændlˌaɪən] *n.* [植] 蒲公英

deamination [diːˌæmiˈneiʃən] *n.* 去氨基；脱氨基作用

debate [dɪˈbeɪt] *v.* 争辩，争论，辩论；*n.* 争论，争议，辩论会，议论

decompose [ˌdikəmˈpoz] *vt.* & *vi.* 分解；(使) 腐烂

defecate [ˈdefəkeɪt] *v.* 澄清 *vt.* 排便

deficiency [dɪˈfɪʃənsɪ] *n.* 缺陷，缺点；缺乏；不足的数额

define [dɪˈfaɪn] *v.* 规定；使明确；(给词、短语等) 下定义

deformity [dɪˈfɔrməti] *n.* 畸形；残废

dehydrated alfalfa meal 脱水苜蓿粉

dehydration [ˌdihaɪˈdreʃən] *n.* 脱水；失水；干燥，极度口渴

dehydrogenate [diːhaɪˈdrɒdʒəneɪt] *vt.* 使脱氢

delicious [dɪˈlɪʃəs] *adj.* 美味的；可口的

denature [diːˈneɪtʃə] *vt.* 使改变本性；使变质

dendrite [ˈdendraɪt] *n.* 枝状突起；树突

dendritic [denˈdritik] *adj.* 树枝状的

depression [dɪˈprɛʃən] *n.* 萎靡不振，沮丧；下陷处，坑；衰弱；减缓

determine [dɪˈtɜːmɪn] *v.* (使) 下决心，(使) 做出决定 *vt.* 决定，确定；判定，判决；限定 *vi.* 确定；决定；判决，终止

development [dɪˈvɛləpmənt] *n.* 发展，进化；被发展的状态；新生事物，新产品；发育

diabetes [daɪəˈbitiz] *n.* 糖尿病

diametrically [ˌdaɪəˈmɛtrɪkli] *adv.* 完全地；作为直径地；直接地；正好相反地

diarrhoea [ˌdaɪəˈrɪə] *n.* 腹泻

dietary [ˈdaɪˌtɛri] *adj.* 饮食的，规定食物的

digest [daɪˈdʒest] *vt.* 消化；吸收；融会贯通；*vi.* 消化；*n.* 文摘；摘要

digestibility [daɪdʒɛstəˈbɪləti] *n.* 消化性；可消化性；消化率

digestive system 消化系统

diligent [ˈdɪlɪdʒənt] *adj.* 勤勉的；用功的，费尽心血的

diphtheria [dɪfˈθɪərɪə] *n.* 白喉

disaccharide [daɪˈsækəraɪd] *n.* 二糖

discipline [ˈdɪsəplɪn] *n.* 纪律；训练；学科 *vt.* 训练；惩罚

disease control 疾病控制

distortion [dɪsˈtɔrʃən] *n.* 扭曲，变形；失真，畸变；扭转

distribution [ˈdɪstrəˈbjʊʃən] *n.* 分布；分发；分配；散布；销售量

division [dɪ'vɪʒn] *n.* 分裂，部门，除法，部类

docile ['dɑːsl] *adj.* 温顺的；驯服的；易驾驭的；驯化

domestic [də'mɛstɪk] *adj.* 家庭的；国内的；驯养的

domesticate [də'mɛstɪket] *vt.* 驯养；教化；引进；*vi.* 驯养；使习惯于或喜爱家务和家庭生活

donors [donər] *n.* 供体，施主，捐赠者，供血者，给予体

dormant ['dɔːrmənt] *adj.* 潜伏的，蛰服的，休眠的；静止的；

dry matter 干物质

dysfunction 功能障碍

E

earthworm ['ɜːθwɜːm] *n.* 蚯蚓

ecology [ɪ'kɑlədʒi] *n.* 生态学；生态

edible ['ɛdəbl] *adj.* 可食用的 *n.* 食品；食物

ejaculation [ɪˌdʒækjʊ'leɪʃən] *n.* 射精

elasticity [ilæ'stisiti] *n.* 弹性，弹力，灵活性

electrofusion [elektrɒf'juːʒn] *n.* 电熔，电融合

element ['elɪmənt] *n.* 元素；要素；原理；成分；自然环境

emaciation [ɪˌmeʃɪ'eʃən] *n.* 消瘦，憔悴，衰弱

embryo [ɛmbrɪo] *n.* 胚胎，胎儿

emphasize ['ɛmfəˌsaɪz] *v.* 强调，着重；使突出

enclosure [ɪn'kloʒər] *n.* 附件；围墙；围场

endocrine gland 内分泌腺

endocrine system 内分泌系统

engage [ɪn'gedʒ] *vt.* 吸引，占用；使参加；雇佣；使订婚；预订 *vi.* 从事；答应，保证；交战；啮合

engulf [ɪn'gʌlf] *vt.* 吞没；吞食=ingulf

enlighten [ɪn'laɪtn] *v.* 启发；开导；教导

entail [ɪn'teɪl] *v.* 需要；牵涉；使必要；限定继承

enteropathogen 肠病原体

enterotoxemia [ˌentərəʊtɒk'siːmɪə] *n.* （羊的）肠毒血病

enthusiasm [ɪn'θjuːziæzəm] *n.* 热情，热忱；热衷的事物；宗教的狂热

environment [ɪn'vaɪrənmənt] *n.* 环境，外界；周围，围绕；工作平台；（运行）环境

environmental factor 环境因素

enzyme [ˈenzaɪm] *n.* 酶
epigenetic [epɪdʒɪˈnetɪk] *n.* 表观遗传学
epinephrine [ˌepɪˈnefrɪn] *n.* 肾上腺素
epithelial tissue 皮组织
epithelium [ˌepɪˈθiːlɪəm] *n.* 上皮，上皮细胞
equivalent [ɪˈkwɪvələnt] *adj.* 相等的，相当的，等效的；等价的 *n.* 对等物
erythrocyte [ɪˈrɪθrəsaɪt] *n.* 红细胞，红血球
Escherichia coli 大肠杆菌
essential fatty acid 必需脂肪酸
ester [ˈestə] *n.* 酯
ether [ˈiːθə] *n.* 乙醚；以太；苍天；天空醚
ether extract 粗脂肪
ethology [iːˈθɒlədʒɪ] *n.* 动物行动学，道德体系学
etiology [ˌiːtɪˈɒlədʒɪ] *n.* 病因学
euthanize [ˈjuːθənaɪz] *vt.* 使安乐死，对……施无痛致死术
evaluation [ɪˌvæljʊˈeʃən] *n.* 评估
eventually [ɪˈventʃuəli] *adv.* 终于，终究，竟
ewe [ju] *n.* 母羊
excel [ɪkˈsɛl] *v.* 优于，擅长
excellent-quality 优质的
excitability 兴奋性
exclusively [ɪkˈsklusɪvlɪ] *adv.* 唯一地；特定地；
excretion [ɪkˈskriːʃən] *n.* 排泄，排泄物；分泌，分泌物
exert [ɪgˈzɜːt] *vt.* 运用，发挥；施与影响
exhaustion [ɪgˈzɔstʃən] *n.* 疲惫，衰竭；枯竭，用尽；排空；彻底的研究
exocrine [ˈɛksəkrɪn] *adj.* 外分泌的
expertise [ˈɛkspɜːˈtiz] *n.* 专门知识；专门技术；专家的意见
exploitation [ˌɛksplɔɪˈteʃən] *n.* 开发；开采；剥削；利用
extend [ɪkˈstɛnd] *v.* 延伸；扩大；推广
extensive [ɪkˈstɛnsɪv] *adj.* 广阔的，广大的
external supply 外源供应
extract [ɛkstrækt] *v.* 提取，抽出，开采，摘录

F

facilitate [fəˈsɪlɪteɪt] *vt.* 帮助，使容易，促进

farm advisor 农场顾问

farm animal 家畜

fat [fæt] *adj.* 肥的，胖的；油腻的；丰满的；*n.* 脂肪，肥肉；*vt.* 养肥；在……中加入脂肪；*vi.* 长肥

fatal [ˈfeɪtəl] *adj.* 致命的；重大的；毁灭性的；命中注定的

FCT Fibrous Connective Tissue 纤维结缔组织

feed [fiːd] *vt.* 喂养；供给；放牧；抚养（家庭等）；靠……为生；*vi.* 吃东西；流入；*n.* 饲料；饲养；（动物或婴儿的）一餐

feeding value 饲喂价值

feedstuff [ˈfiːdstʌf] *n.* 饲料（等于 feedingstuff）

felid [ˈfiːlɪd] *n.* 猫科动物

fencelike [ˈfenslaik] *adj.* 像栅的，像栅栏的，像围栏的，像篱笆的，像围墙的

feral [ˈferəl] *adj.* 野生的

fermentation product 发酵产物

fertilization [ˌfɜːtləˈzeʃən] *n.* 施肥；受精，受精过程，受精行为；受孕；受胎

fetal [ˈfiːtl] *adj.* 胎儿的，胎的

fetus [ˈfiːtəs] *n.* 胎儿，胎

fibroblast [ˈfaɪbrəʊblæst] *n.* 纤维原细胞，成纤维细胞

flavor [ˈflevər] *n.* 味；韵味；特点；香料 *vt.* 给……调味；给……增添风趣

flesh [flɛʃ] *n.* 肉；肉体；果肉；皮肤 *v.* 用肉喂养；长胖

flexibility [fleksɪˈ biliti] *n.* 柔性，灵活性

flighty [ˈflaɪti] *adj.* 反复无常的

flock [flɒk] *n.* 兽群，鸟群；群众；棉束；大堆，大量 *vi.* 群集，成群结队而行 *vt.* 用棉束填

fluorine [ˈflʊəriːn] *n.* 氟

foal [fəʊl] *n.* 驹（尤指一岁以下的马、驴、骡）；*vi.*（马等）生仔；*vt.*（马等）生仔

focus [ˈfəʊkəs] *n.* 疫源地

foodstuff [ˈfudstʌf] *n.* 食品；食料

foot-and-mouth disease 口蹄疫

forage [ˈfɒrɪdʒ] *n.* 饲料；草料；搜索；*vi.* 搜寻粮草；搜寻

foster [ˈfɒstər] *v.* 培养；促进；抚育；代养 *adj.* 寄养的；代养的

frog [frɔːg] *n.* 青蛙
frost [frɔst] *n.* 霜冻，严寒天气
fructan [ˈfrʌktən] *n.* 果聚糖
fructose [ˈfrʌktəʊz] *n.* 果糖；左旋糖
frustrate [frʌˈstreɪt] *vt.* 挫败；阻挠；使受挫折 *adj.* 无益的，无效的
fumaric acid 延胡索酸
furnish [ˈfərnɪʃ] *vt.* 陈设，布置；提供，供应；装修（房屋）

G

galactose [gəˈlæktəʊz] *n.* 半乳糖
gametogenesis [gəˌmitəˈdʒɛnɪsɪs] *n.* 配子
gastrointestinal tract 消化道
genetic improvement 基因改良
genetic information 遗传信息
genetic manipulation：遗传调控
geographic [dʒiːəˈgræfɪk] *adj.* 地理学的，地理的 .
geographical [dʒɪəˈgræfɪkəl] *adj.* 地理的；地理学的
gestation [dʒɛˈsteʃən] *n.* 妊娠期
glanders [ˈglændəz] *n.* 鼻疽
glial cell 神经胶质细胞
glucose [ˈgluːkəʊs] *n.* 葡萄糖；葡糖
glycerol [ˈglɪsərɒl] *n.* 甘油；丙三醇
glycine [ˈglaɪsiːn] *n.* 甘氨酸；氨基乙酸
glycogen [ˈglaɪkədʒən] *n.* 糖原；动物淀粉
goblet cells 杯状细胞
grass [grɑːs] *n.* 草；草地，草坪；*vt.* 放牧；使……长满草；使……吃草；*vi.* 长草
grazing [ˈgreɪzɪŋ] *n.* 放牧；牧草；*v.* 擦过；抓伤（graze 的现在分词）
groin [grɔɪn] *n.* 腹股沟
groundnut [ˈgraʊndnʌt] *n.* 落花生；野豆
gut [gʌt] *n.* 内脏

H

hairball [ˈherˌbɔːl] *n.*（动物在胃或肠积成的）毛团
halal [ˈhælæl] *adj.* 伊斯兰教律法允许的食物等；按伊斯兰教律法售卖或

提供食物的

handler [ˈhændlə] *n.* 处理者；管理者；（犬马等的）训练者

Hantaan virus 汉坦病毒

harbor [ˈhɑrbə] *n.* 港口；*v.* 隐藏，庇护，藏匿

hatchery [ˈhætʃəri] *n.*（尤指鱼的）孵化场

hay [heɪ] *n.* 待割的草；干草

heifer [ˈhefə] *n.* 小母牛

hemoglobin 血红蛋白

hemorrhagic fever 出血热

hence [hɛns] *adv.* 从此；因此，所以

hepatitis B B 型肝炎

herd [hɜːd] *n.* 兽群，畜群；放牧人；*vi.* 成群，聚在一起；*vt.* 放牧；使成群

heredity [həˈrɛdəti] *n.* 遗传；遗传特征

heterozygotes [ˌhetərəˈzaɪgoʊt] *n.* 杂合子

hexose [ˈheksəʊz] *n.* 己糖

hexose phosphates 磷酸己糖

histidine [ˈhɪstɪdiːn] *n.* 组氨酸

hogs [hogz] *n.* 育肥猪

holding area 驯养区

homeostatic [ˌhoʊmɪrˈsteɪtɪk] *adj.*（社会群体的）自我平衡的，原状稳定的

homologous [həˈmɒləgəs] *adj.* 同源的；相应的；类似的；一致的

homozygous [ˌhɒməˈzaɪgəs] *n.* 纯合子

hooves [hʊvzˌhuvz] *n.* hoof 的复数；（兽的）蹄，马蹄 hoof 的名词复数

horse-breeding farm 繁育马场

humidity [hjuˈmɪdəti] *n.* 湿度；湿气

humoral immunity 体液免疫

husbandry [hʌzbəndri] *n.* 畜牧业，饲养业

hydrolysis [haɪˈdrɒlɪsɪs] *n.* 水解作用

hydroxyproline [haiˌdrɔksiˈprəuliːn] *n.* 羟（基）脯氨酸

hypersensitivity [ˌhaɪpərˌsɛnsəˈtɪvɪti] *n.* 超敏性

hypothermia [ˌhaɪpəˈθɜːmiə] *n.* 低体温

I

ideally [aɪˈdiəli] *adv.* 理想地

ignition [ɪgˈnɪʃən] *n.* 点火，点燃；着火，燃烧；点火开关，点火装置

immune [ɪˈmjuːn] *adj.* 免疫的

immune response 免疫反应

immune system 免疫系统

immunocompromising serovar 血清型

immunodeficiency [ˌɪmjʊnəʊdɪˈfɪʃənsɪ] *n.* 免疫缺陷

immunoglobulin [ɪˈmjuːnəʊˈglɒbjʊlɪn] *n.* 免疫血球素，免疫球蛋白

immunologic memory 免疫记忆

immunosuppression [ˌɪmjunosəˈprɛʃən] *n.* 〈医〉免疫抑制

impact [ɪmˌpækt] *n.* 影响；碰撞；*vt.* 撞击；挤入，压紧；对……产生影响；*vi.* 产生影响；冲撞.

impinge [ɪmˈpɪndʒ] *vi.* 撞击；侵犯；对……有影响 *vt.* 撞击，打击

implicate [ˈɪmplɪkeɪt] *vt.* 暗示；牵涉，涉及（某人）；表明（或意指）……是起因 *n.* 包含的东西

improve [ɪmˈpruv] *vt.* 提高（土地、地产）的价值；利用（机会）；改善，改良 *vi.* 变得更好；改进，改善

inactivate [ɪnˈæktɪveɪt] *vt.* 使不活动；使变不活泼；去激活

inbreeding [ˈɪnbriːdɪŋ] *n.* 近亲交配；同系繁殖

incorporate [ɪnˈkɔːpəreɪt] *vi.* 合并；包含；吸收；混合

incremental [ɪnkrəˈməntl] *adj.* 增加的，增值的

independently [ɪndɪˈpɛndəntli] *adv.* 独立地

indispensable [ˈɪndɪˈspɛnsəbl] *adj.* 不可缺少的；绝对必要的；责无旁贷的 *n.* 不可缺少之物；必不可少的人

inevitably [ɪnˈevɪtəbli] *adv.* 难免；不可避免地

infection [ɪnˈfekʃən] *n.* 感染；传染；影响；传染病

infestation [ˌɪnfeˈsteɪʃən] *n.* 感染；侵扰

inflate [ɪnˈfleɪt] *vt.* & *vi.* 使充气（于轮胎、气球等）；（使）膨胀

inflict [ɪnˈflɪkt] *vt.* 造成；使遭受（损伤、痛苦等）；给予（打击等）

ingredients [ɪŋˈgriːdɪrnts] *n.* （烹调的）原料；（构成）要素；因素

innumerable [ɪˈnuːmərəbl] *adj.* 无数的，不计其数，不可胜数

Inseminating Rod [ɪnˈsemɪneɪt] [rɑːd] *n.* 输精枪；

instinct [ˈɪnˌstɪŋkt] *n.* 本能，天性；冲动；*adj.* 深深地充满着

instruction [ɪnˈstrʌkʃən] *n.* 指令，命令；指示；教导；用法说明

insufficient [ˌɪnsəˈfɪʃnt] *adj.* 不足的，不够的

insulation [ɪnsjʊˈleɪʃən] *n.* 绝缘，隔离，孤立，绝缘或隔热的材料，隔声

insulin [ˈɪnsjʊlɪn] *n.* 胰岛素
intensive [ɪnˈtɛnsɪv] *adj.* 加强的，强烈的
interaction [ˌɪntəˈrækʃən] *n.* 一起活动；合作；互相影响；互动
intercalated disks 闰盘
intercede [ˌɪntəˈsiːd] *vi.* 说情；斡旋；调解
interfere [ˌɪntəˈfɪər]；*vi.* 干预，干涩；调停，排解；妨碍，干扰
intermediate [ˌɪntəˈmiːdɪət] *n.* 媒介
intermediate product 中间产物
interruption [ɪntəˈrʌpʃən] *n.* 中断；打断；障碍物；打岔的事；
intervention [ˌɪntəˈvenʃn] *n.* 介入，干涉，干预；调解，排解
intestinal [ɪnteˈstaɪnəl] *n./adj.* 肠的
intestine [inˈtestin] *n.* 肠
intracellular [ˌɪntrəˈseljʊlə] *adj.* 细胞内的
invade [ɪnˈveɪd] *vt.* & *vi.* 侵入，侵略；进行侵略；蜂拥而入，挤满；（疾病，声音等）袭来，侵袭 *vt.* 侵犯；侵袭；涌入；干扰
invader [ɪnˈveɪdər] *n.* 入侵物，入侵者
invasive [ɪnˈvesɪv] *adj.* 侵略性的，侵害的；攻击性的；扩散性的，蔓延性的
iron [ˈaɪən] *n.* 熨斗；烙铁；坚强；*vt.* 熨；用铁铸成；*adj.* 铁的；残酷的；刚强的；*vi.* 熨衣；烫平
irrigation [ɪrɪˈgeʃn] *n.* 灌溉；冲洗；冲洗法
isinglass [ˈaɪzɪŋglæs] *n.* 鱼胶；明胶；云母
isoleucine [ˌaɪsəʊˈluːsiːn] *n.* 异亮氨酸

K

keratin [ˈkerətin] *n.* 角蛋白
keto acid 酮酸
ketobutyric acid 酮基丁酸
ketosis [kɪˈtəʊsɪs] *n.* 酮病
kidney [ˈkɪdnɪ] *n.* 肾脏；腰子；个性
kidney tubules 肾小管
kosher [ˈkoʃər] *n.* 清洁可食的食物 *v.* 使（食物）清洁可食

L

labour [ˈlebər] *n.* 分娩

lactating dairy cow 泌乳牛

lactation [læk'teɪʃən] n. 哺乳；哺乳期；授乳（形容词 lactational）；分泌乳汁

lactic acid 乳酸

lactose ['læktəʊz] n. 乳糖

ladino clover 白三叶草

lamb [læm] n. 羔羊；小羊

lambing ['læmɪŋ] adj. 产羔羊的；n. 产羔羊

lameness [leɪmnəs] n. 跛行；跛，残废，僵而疼痛

laminitis [ˌlæmə'naɪtɪs] n. 蹄叶炎

LOS large offspring syndrome 大肽综合征

latent ['leɪtənt] adj. 潜伏的

laxative ['læksətɪv] n. 泻药；缓泻药

LCT Loose Connective Tissue 疏松结缔组织

lecithin ['lesɪθɪn] n. 卵磷脂；蛋黄素

legislator ['lɛdʒɪsletər] n. 立法委员；立法者

legume ['legjuːm] n. 豆类；豆科植物；豆荚

length of daylight 日照时间

leptospirosis [ˌleptəʊspaɪ'rəʊsɪs] n. 钩端螺旋体病

lesion ['liːʒən] n. 损害，病变

leucine ['luːsiːn] n. 亮氨酸；白氨酸

leukocyte ['ljuːkəʊsaɪt] n. 白细胞，白血球

ligament ['lɪgəmənt] n. 韧带

likewise ['laɪkˌwaɪz] adv. 同样地；也，而且

lipid ['lɪpɪd] n. 脂质；油脂

listeria monocytogenes 单核细胞增多性李斯特菌

literally ['lɪtərəli]；adv. 逐字地；照字面地

litter ['lɪtər] n. 杂物，垃圾；（一窝）幼崽；褥草；轿，担架 vt. & vi. 乱扔；使杂乱；乱丢杂物；使饱含

liver ['lɪvə] n. 肝脏；生活者，居民

livestock ['laɪvstɑk] n. 牲畜；家畜

longevity [lɔn'dʒɛvəti] n. 长寿，长命；寿命

lose sight of 看不到，看不见；失去与……的联系；忘记；忽略

louping-ill virus 羊传染性脑脊髓炎病毒

lubricant ['luːbrɪkənt] n. 润滑剂，润滑油 adj. 润滑的

lush [lʌʃ] *adj.* 丰富的，豪华的；苍翠繁茂的；*vi.* 喝酒；*n.* 酒；酒鬼；*vt.* 饮

lymph node 淋巴结

lymphocyte [ˈlɪmfəʊsaɪt] *n.* 淋巴球，淋巴细胞

lysine [ˈlaɪsiːn] *n.* 赖氨酸

M

macrophage [ˈmækrəʊfeɪdʒ] *n.* 巨噬细胞

maggot [ˈmægət] *n.* 蛆；*adj.* 多蛆的

malic acid 苹果酸

mammary [ˈmæməri] *n.* 乳房的，乳腺的

mammary gland 乳腺

manipulate [məˈnɪpjulet] *vt.* 操纵；操作；巧妙地处理；篡改

mannose [ˈmænəʊz] *n.* 甘露糖

manure [məˈnjʊər] *n.* 肥料，粪便

manure [məˈnʊr] *n.* 肥料 *vt.* 施肥

manure [məˈnʊr] *vt.* 施肥于；耕种 *n.* 肥料；粪肥

massage [ˈmæsɑːʒ] *n.* 按摩，推拿

matrix [ˈmeɪtrɪks] *n.* 基质

mature [məˈtjuə] *vi.* 成熟；*adj.* 成熟的

mature [məˈtʃʊə] *adj.* 成熟的；充分考虑的；到期的；成年人的；*vi.* 成熟；到期；*vt.* 使……成熟；使……长成；慎重做出

measles [ˈmiːzəlz] *n.* 麻疹

meconium [məˈkəʊnɪəm] *n.* 胎尿，胎粪

membrane [ˈmembreɪn] *n.* 膜；薄膜；羊皮纸

metabolism [mɪˈtæbəlɪzəm] *n.* 新陈代谢

methionine [mɪˈθaɪəniːn] *n.* 蛋氨酸

metronome [ˈmetrənəʊm] *n.* 节拍器

microbe [ˈmaɪkrəʊbz] *n.* 微生物，细菌

microbiota [maɪkrəʊbaˈɪɒtə] 微生物丛，微生物区

milking herd 挤奶牛群

mineral [ˈmɪnərəl] *n.* 矿物；（英）矿泉水；无机物；苏打水（常用复数表示）；*adj.* 矿物的；矿质的

minister [ˈmɪnɪstər] *n.* 部长；大臣；牧师；*vi.* 执行牧师职务；辅助或伺候某人

minute [ˈmɪnɪt] *n.* 分，分钟；片刻，一会儿；备忘录，笔记；会议记录；*vt.* 将……记录下来；*adj.* 微小的，详细的

moderate [ˈmɒdərət] *adj.* 温和的；适度的，中等的

modify [ˈmɑdɪfaɪ] *vt.* 修改，修饰；更改 *vi.* 修改

molecular [məˈlekjələr] *adj.* 分子的，由分子组成的

molecular weight 分子量

monogastric animal 单胃动物

monosaccharide [mɒnəʊˈsækəraɪd] *n.* 单糖，单糖类（最简单的糖类）

morbilli virus 麻疹病毒

mucosa [mjuːˈkəʊsə] *n.* 黏膜

mucus [ˈmjuːkəs] *n.* 黏液，胶

multinucleate [ˌmʌtiˈnjuːkliit] *adj.* 多核的

muscle tissue 肌肉组织

mycotoxin [ˌmaɪkoʊˈtɒksən] *n.* 霉菌毒素

myofibril [ˌmaɪəˈfaɪbrəl] *n.* 肌原纤维

myosin [ˈmaɪəʊsɪn] *n.* 肌浆球蛋白，阻凝蛋白

N

nail [neɪl] *vt.* 钉；使固定；揭露；*n.* [解剖] 指甲；钉子

nervous tissue 神经组织

nervous tissue 神经组织

neuron [ˈnjʊərɒn] *n.* 神经细胞，神经元

neuroscientists [ˌnjʊərəʊˈsaɪəntɪst] *n.* 神经系统科学家

neuter [ˈnuːtər] *adj.* 中性的，不及物的，（生物）无性的 *n.* 中性名词，无性动物，阉割动物

niche [niːʃ] *n.* 合适的位置（工作等）；有利可图的缺口，商机

nitrate [ˈnaɪˌtretˌ，-trɪt] *n.* 硝酸盐；硝酸根；硝酸酯；硝酸盐类化肥

nitrogen-containing compound 含氮化合物

nitrogen-containing organic compound 含氮有机化合物

nitrogen-free extractive 无氮浸出物

non-ambulatory [ˈæmbjələtɔːri] *adj.* 非走动的，非流动的

nonspecific defense 非特异性防疫系统

notwithstanding [ˌnɑtwɪθˈstændɪŋ] prep. 尽管；虽然 *adv.* 尽管如此，仍然；还是 *conj.* 虽然，尽管

noxious [ˈnɑkʃəs] *adj.* 有害的；有毒的；败坏道德的；讨厌的

nucleic acid 核酸

nucleus 细胞核

nudge [nʌdʒ] *vt.* 推进；（用肘）轻推

nurse [nɜːs] *n.* 哺乳，吃乳；

nurture [ˈnɜːtʃə] *n.* 养育，培养

nutrient [ˈnjuːtrɪənt] *n.* 营养物；滋养物；*adj.* 营养的；滋养的

nutritional value 营养价值

nutritive [ˈnjʊtrətɪv] *adj.* 有营养的；滋养的；有营养成分的；与营养有关的 *n.* 营养物

O

oat [əʊt] *n.* 燕麦；麦片粥，燕麦粥

occasionally [əˈkeɪʒnəli] *adv.* 偶尔，偶然，有时候

odor [ˈoʊdə] *n.* 气味；名声；气息

offspring [ɔfsprɪŋ] *n.* 子孙，后裔

oilseed [ˈɒɪlsiːd] *n.* 含油种子（如花生仁、棉籽等）

oocytes [uːsaɪts] *n.* 卵母细胞

ooplasm [oʊəplæzm] *n.* 卵胞质

operation [ˌɑːpəˈreɪʃn] *n.* 操作，经营

opportunity [ˌɑːpərˈtuːnəti] *n.* 机会

opt [ɒpt] *vi.* 选择，挑选

optimum [ˈɑːptɪməm] *adj.* 最适宜的

orchardgrass [ˈɔːrtʃərˈgræs] 野茅；果园草

organ [ɔrgən] *n.* 器官

organic acid 有机酸

organism [ˈɔːgənɪzəm] *n.* 有机体；生物体；微生物；有机体系，有机组织

osteocyte [ˈɒstɪəʊˌsaɪt] *n.* 骨细胞

overgraze [ˌovərˈgrez] *vi.* 过度放牧；*vt.* 在……上过度放牧

oxidation [ɒksɪˈdeɪʃən] *n.* 氧化

oxygen [ˈɒksɪdʒən] *n.* 氧气，氧

P

parasite [ˈpærəsaɪt] *n.* 寄生虫；食客

parasitology [ˌpærəsɪˈtɒlədʒɪ] *n.* 寄生虫学

parsley [ˈpɑrsli] *n.* 西芹，欧芹；洋芫荽

Parturition [ˌpɑːtjʊˈrɪʃən] *n.* 分娩，生产

pasture [ˈpæstʃər] *n.* 牧场；草原 *vi.* 吃草 *vt.* 放牧

pathogen [ˈpæθədʒən] *n.* 病菌，病原体

pathogenic *adj.* 引起疾病的

pathology [pəˈθɒlədʒɪ] *n.* 病理学

pelvis [ˈpɛlvɪs] *n.* 骨盆

pentose [ˈpentəʊz] *n.* 戊糖

peptone [ˈpeptəʊn] *n.* 蛋白胨，胨

perception [pərˈsepʃn] *n.* 知觉；观念

periodically [ˌpɪərɪˈɒdɪkəlɪ] *adv.* 定期地；周期性地；偶尔；间歇

peripheral [pəˈrɪfərəl] *adj.* 外围的；次要的；（神经）末梢区域的；*n.* 外部设备

perished [ˈpɛrɪʃt] *adj.* 感觉很冷的；脆裂的 *v.* 灭亡（perish 的过去分词）；枯萎

permeability [pɜːmɪəˈbɪlɪtɪ] *n.* 渗透性；透磁率，导磁系数；弥漫

phase [feɪz] *n.* 时期，阶段，局面，段落，学时

phenomena [fəˈnɒmɪnə] *n.* 现象，phenomenon 的复数形式

phenotypic [ˌfiːnəʊˈtɪpɪk] *n.* 表型

phenylalanine [ˌfiːnaɪlˈæləniːn] *n.* 苯基丙氨酸

philosophical [ˌfɪləˈsɑfɪkl] *adj.* 哲学上的，哲学（家）的；冷静的，沉着的；明达的；达观的

phosphate [ˈfɒsfeɪt] *n.* 磷酸盐；皮膜化成

phosphorus [ˈfɒsfərəs] *n.* 磷

photosynthesis [ˌfəʊtəʊˈsɪnθɪsɪs] *n.* 光合作用

physical [ˈfɪzɪkl] *adj.* 物理的；身体的；物质的；根据自然规律的，符合自然法则的；*n.* 体格检查

physiological [ˌfɪzɪəˈlɒdʒɪkəl] *adj.* 生理学的，生理的

plague [pleɪg] *n.* 鼠疫

plasma [ˈplæzmə] *n.* 等离子体；血浆；深绿玉髓

platelet [ˈpleitlit] *n.* 血小板

pleuropneumonia [ˌplʊərəʊnjuːˈməʊnɪə] *n.* 胸膜性肺炎

polygenic [ˌpɑlɪˈdʒenɪk] *adj.* 多基因的（遗传特征）

polysaccharide [ˌpɒlɪˈsækəraɪd] *n.* 多糖；多聚糖

populate [ˈpɒpjuleɪt] *vt.* 居住于；生活于；移民于；落户于

potassium [pəˈtæsɪəm] *n.* 钾

potential [pəˈtɛnʃəl] *adj.* 潜在的

pouch [paʊtʃ] *n.* 小袋，育儿袋

predisposition [ˌpridɪspəˈzɪʃən] *n.* 倾向，素质；易染病体质

preparation *n.* 制剂，配制品

presence [ˈprɛzəns] *n.* 出席；仪表；风度

presumably [prɪˈzuːməbli] *adv.* 大概；可能

primitive [ˈprɪmətɪv] *adj.* 原始的；简陋的 *n.* 文艺复兴前的艺术家；原始人

principally [ˈprɪnsəpli] *adv.* 主要地；大部分

probiotic [ˌprobaɪˈɑtɪk] *n.* 益生菌

procedures [prəˈsiːdʒəz] *n.* 程序；手续；手续；步骤；常规

process [ˈproʊses] *n.* 过程 *v.* 处理

productivity [prɑːdʌkˈtɪvəti] *n.* 生产率，生产力

profitability [ˌprɑfɪtəˈbɪlətɪ] *n.* 获利（状况），盈利（情况）

profitable [ˈprɒfɪtəbəl] *adj.* 有利可图的；赚钱的；有益的

prognosis [prɒgˈnəʊsɪs] *n.* 预断病情；预后；预测

proline [ˈprəʊliːn] *n.* 脯氨酸

prolonged [prəˈlɔːŋd] *adj.* 延长的；持续很久的；拖延的

propagation [ˌprɑpəˈgeʃən] *n.* 传播；繁殖；增殖

propionic acid 丙酸

proponent [prəˈpoʊnənt] *n.* 提倡者；支持者

proportion [prəˈpɔːrʃn] *n.* 比例

protein [ˈprəʊtiːn] *n.* 蛋白质；朊；*adj.* 蛋白质的

proteose [ˈprəʊtɪəʊs] *n.* 朊间质；蛋白脲；（蛋白）脲

provision [prəˈvɪʒən] *n.* 规定，条项，条款；预备，准备，设备；供应，（一批）供应品；生活物质，储备物资

prussic acid [ˈprʌsɪk ˈæsɪd] *n.* 氰酸，氢氰酸

psychological [ˌsaɪkəˈlɑdʒɪkl] *adj.* 心理的；心理学的；精神上的

purchase [ˈpɜːrtʃəs] *v.* 购买

pyruvate [paɪˈruːveɪt] *n.* 丙酮酸盐；丙酮酸酯

pyruvic acid 丙酮酸

Q

quality [ˈkwɒlətɪ] *n.* 质量，品质；特性；才能；复数 qualities

quantity [ˈkwɒntɪtɪ] *n.* 量，数量；大量；总量；复数 quantities

quarters [ˈkwɔtərz] *n.* 住处，岗位

R

ranges [reɪndʒ] *n.* 范围；围栏牧场；射程；山脉；排；一系列；闲逛；炉灶 *v.* 排列；使……站在某一方；延伸；漫游

reaction [rɪˈækʃən] *n.* 反应，感应；反动，复古；反作用

recipient [rɪˈsɪpiənt] *n.* 接受者；容器；容纳者；*adj.* 容易接受的；感受性强的

recommendation [rɛkəmɛnˈdeʃən] *n.* 推荐；建议；推荐信；可取之处

rectify [ˈrektɪfaɪ] *vt.* 改正，校正

rectum [rɛktəm] *n.* 直肠

red blood cell 红血球

red clover 红三叶草

refer [rɪˈfɜːr] *v.* 简称，参考，参照，谈到，提交

regime [reˈʒimˌrɪ-] *n.* 管理，方法；(病人等的) 生活规则

render [ˈrɛndər] *vt.* 提供；表现；使成为；宣布；翻译；回报；给予补偿；渲染，*n.* 粉刷；打底；交纳

renewable [rɪˈnjuːəbəl] *adj.* 可再生的；可更新的；可继续的；*n.* 再生性能源

rennet [ˈrɛnɪt] *n.* 牛犊胃内膜；凝乳

replacement [rɪˈplesmənt] *n.* 代替

reproductive tract 生殖管

requirement [rɪˈkwaɪəmənt] *n.* 要求；必要条件；必需品

reservoir [ˈrezəvwɑ：r] *n.* 水库，蓄水池

restraint [rɪˈstrent] *n.* 抑制，克制；约束

restriction [rɪˈstrɪkʃən] *n.* 限制；约束；束缚

rhythm [ˈrɪðəm] *n.* 节奏；韵律

rinderpest [ˈrɪndəpest] *n.* 牛瘟

ritual [ˈrɪtʃuəl] *n.* 仪式；惯例；礼制 *adj.* 仪式的；例行的；礼节性的

rotate [rəʊˈteɪt] *vi.* 旋转；循环；*vt.* 使旋转；使转动；使轮流；*adj.* [植] 辐状的

roughages [ˈrʌfɪdʒ] *n.* 粗饲料，粗粮，粗糙的原料

rumen [ˈruːmen] *n.* 瘤胃 (反刍动物的第一胃)

rumen [ruːmen] *n.* 瘤胃反刍动物的第一胃

ruminant animal 反刍动物

rye [raɪ] *n*. 黑麦；吉卜赛绅士；*adj*. 用黑麦制成的

ryegrass [ˈraɪgræs] *n*. 黑麦草

S

s. typhimurium 鼠伤寒沙门菌

salamander [sæləmændər] *n*. 蝾

salmonella [ˌsælməˈnelə] *n*. 沙门菌

sanitary [ˈsænətri] *adj*. 卫生的；清洁的 *n*. 公共厕所

sanitize [ˈsænɪtaɪz] *vt*. 净化；进行消毒；使清洁；审查

saponification [səˌpɒnɪfɪˈkeɪʃən] *n*. 皂化

sarcomere [ˈsɑːkəʊmɪə] *n*. 肌原纤维节，肌小节

sceptical [ˈskeptɪkl] *adj*. 怀疑的；怀疑论者的

scheme [skim] *n*. 计划

sea urchins [si] [ɜrtʃɪn] 海胆

secrete [siˈkriːt] *vt*. 分泌

section [ˈsekʃənz] *n*. 节；部门；部分；部件

sensation [senˈseɪʃn] *n*. 感觉；轰动；激动；知觉

sensitive [ˈsensətɪv] *adj*. 敏感的；感觉的

sentiment [ˈsɛntɪmənt] *n*. 感情，情绪；情操；观点；多愁善感

serine [ˈsɪəriːn] *n*. 丝氨酸

serotype [ˈsɪərətaɪp] *n*. 血清型 *v*. 按血清型分类，决定……的血清型，把……按血清型分类

serum [sɪrəm] *n*. 血清，浆

shelter [ˈʃɛltər] *n*. 庇护；避难所；遮盖物 *vt*. 保护；使掩蔽 *vi*. 躲避，避难

shred [ʃrɛd] *n*. 碎片；破布；少量 *vt*. & *vi*. 撕碎，切碎；用撕毁机撕毁（文件）

significance [sɪgˈnɪfəkəns] *n*. 显著性

silicon [ˈsɪlɪkən] *n*. 硅；硅元素

skeletal [ˈskelɪtəl] *adj*. 骨骼的，像骨骼的；骸骨的；骨瘦如柴的

slaughter [ˈslɔːtə] *vt*. 屠宰，屠杀；杀戮；使惨败；*n*. 屠宰，屠杀；杀戮；消灭

slaughterhouses [ˈslɔːtəhaʊs] *n*. 屠宰场（等于 abattoir）；屠杀场

small intestine 小肠

sodium [ˈsəʊdɪəm] *n.* 钠

soft tissue 软组织

solubility [ˌsɒljʊˈbɪlətɪ] *n.* 溶解度；可解决性

soluble [ˈsɒljʊbəl] *adj.* 可溶的，可溶解的；可解决的

somatic cell nuclear transfer（SCNT）体细胞核移植

somatic cells 体细胞

spay [speɪ] *vt.* 切除卵巢

specific immunity 特异性免疫

specificity [ˌspesɪˈfɪsətɪ] *n.* 特异性

spectrum [ˈspektrəm] *n.* 光谱；波谱；范围；系列

spinach [ˈspɪnɪtʃ] *n.* 菠菜

spleen [spliːn] *n.* 脾

split [splɪt] *v.* 分裂，破裂；*n.* 裂缝，裂开，破裂；*adj.* 分裂的，不一致的

spongy bone 海绵状骨骼

squamous epithelium 扁平上皮

squirt [skwɜːt]；*n.* 喷，细的喷流；注射器

stallion [ˈstæljən] *n.* 种马；成年公马

starch [stɑːtʃ] *n.* 淀粉；刻板，生硬；*vt.* 给……上浆

starvation [stɑːrˈveɪʃn] *n.* 饥饿，饿死

state university 州立大学

stocker cattle 食用牛，候宰牛

stockman [ˈstɑkmən] *n.* 畜牧业者；仓库管理员

stomach [ˈstʌmək] *n.* 胃；腹部；食欲；欲望

strain [stren] *n.* 血统，家族

stratiffed [ˈstrætifaid] *adj.* 成层了的，分层的

strawn [strɔ] *n.* 稻草；吸管；麦秆；毫无价值的东西 *adj.* 稻草的，麦秆的；稻草做的；假的，假想的；无价值的

strep throat 脓毒性咽喉炎

streptococcus 链球菌

suboptimal [ˈsʌbˈɒptɪməl] *adj.* 未达最佳标准的；不最理想的

subsequent [ˈsʌbsɪˌkwɛntˌ-kwənt] *adj.* 随后的；后来的

substantial [səbˈstænʃəl] *adj.* 大量的；实质的；内容充实的；*n.* 本质；重要材料

suburban area 郊区

succinic acid 琥珀酸

succumb [səˈkʌm] *vi.* 屈服；死；被压垮

suffering [ˈsʌfərɪŋ] *n.* 受难；苦楚 *adj.* 受苦的；患病的 *v.* 受苦；蒙受（suffer 的 ing 形式）

sulphur [ˈsʌlfə] *n.* 硫黄；硫黄色；*vt.* 使硫化；用硫黄处理

sunˈsray 太阳光

superfluous [sʊˈpɜ˞flʊəs] *adj.* 过多的；多余的；不必要的；

supervision [sʊpərˈvɪʒən] *n.* 监督；管理；监督的行为，过程或作用

supplement [ˈsʌplɪmənt] *vt.* 增补，补充；*n.* 增补，补充；补充物；增刊，副刊

susceptibility [səˌsɛptəˈbɪlɪti] *n.* 易受影响或损害的状态，感受性

suspension [səˈspenʃn] *n.* 悬浮；悬架；悬浮液；暂停

swab [swɒb] *n.* 药签

symptom [ˈsɪmtəm, ˈsɪmp-] *n.* 症状；征兆

synonymous [sɪˈnɒnɪməs] *adj.* 同义词的；同义的，类义的

synthesis [ˈsɪnθɪsɪs] *n.* 综合，合成；综合体

syphilis [ˈsɪfɪlɪs] *n.* 梅毒

syringe [ˈsɑɪrɪŋ] 注射器，灌注器

T

tarry [ˈtæri] *vi.* 逗留；停留；暂住；徘徊 *adj.* 柏油的；涂柏油的，被柏油弄脏的

teat [tiːt] *n.* 乳头

temperature [temprətʃər] *n.* 温度；量某人的体温；发烧；氛围；体温

temporarily [tempəˈrerɪlɪ] *adv.* 暂时地；临时地

tendency [ˈtendənsɪ] *n.* 倾向，趋势；癖好

tender [ˈtendər]；*adj.* 温柔的；嫩的

tendon [ˈtendən] *n.* 腱，筋

terminal [ˈtɜːrmɪnl] *adj.* 末端的；末期的

testicle [ˈtɛstɪkəl] *n.* 精巢；睾丸

tetanus [ˈtetənəs] *n.* 破伤风

thaw [θɔː] *vt.* 使融化

threonine [ˈθriːəniːn] *n.* 苏氨酸；羟丁胺酸

timothy [ˈtɪməθɪ] *n.* 梯牧草

toxoplasmosis [ˌtɒksəʊplæzˈməʊsɪs] *n.* 弓形虫病

trace mineralized salt 微量矿化盐
trait [tret] *n.* 特点，特性；少许；性状
transmit [trænz'mɪt] *v.* 传递，遗传，传导，播放，发送，发射
transplant [trænsplænt] 移植
tricarboxylic cycle 三羧酸循环
tryptophan ['trɪptəfæn] *n.* 色氨酸
tularemia [ˌtʊlə'rimɪə] *n.* 兔热病
typhoid ['taɪfɒɪd] fever 伤寒
tyrosine ['taɪrəsiːn] *n.* 酪氨酸

U

udder ['ʌdər] *n.* （牛、羊等的）乳房
underfed *adj.* 营养不良的
undergo [ʌndər'goʊ] *v.* 经历，经受，遭受，意会
uniformity [ˌjuːnɪ'fɔːmətɪ] *n.* 一致性；均匀性
unsurpassed [ˌʌnsər'pæst] *adj.* 未被超越的；非常卓越的
upend [ʌp'end] *v.* 颠倒，倒放
urethral [jʊ'riθrəl] *adj.* [解剖] 尿道的
urinary ['jʊrəneri] *adj.* 尿的；泌尿器的；泌尿的 . *n.* 小便池
urinary bladder 膀胱
uterine ['jutərɪn] *adj.* 子宫的，同母异父的，母系的

V

vaccination [ˌvæksɪ'neɪʃən] *n.* 接种疫苗
vagina [və'dʒaɪnə] *n.* 阴道；[植] 叶鞘
vague [veg] *adj.* 模糊的；（思想上）不清楚的；（表达）含糊的
valine ['veɪliːn] *n.* 缬氨酸
vegan ['vigən] *n.* 严格的素食主义者
veganism ['vedʒənɪzəm] *n.* 纯素食主义
vegetarian [ˌvɛdʒə'tɛrɪən] *n.* 素食者；食草动物 *adj.* 素食的
vegetation [ˌvɛdʒɪ'teʃən] *n.* 植物（总称），草木；赘生物，增殖体
venom ['venəm] *n.* 毒液
versus ['vɜːrsəs] prep. 对抗；（比较两种想法、选择等）与……相对
vertebrate ['vɜːtɪˌbreɪt] *adj.* 脊椎动物的；有脊椎的；*n.* 脊椎动物
veterinarian ['vɛtərə'nɛrɪən] *n.* 兽医

vice versa 反之亦然

vitamin [ˈvɪtəmɪn] *n.* 维生素

volatile [ˈvɒlətaɪl] *adj.* 挥发性的；不稳定的；爆炸性的；反复无常的；*n.* 挥发物；有翅的动物

voluntary culling ：主动淘汰

vulva [vʌlvə] *n.* 阴户，女阴；孔

W

waste product 排泄物

wax [waks] *n.* 蜡；蜡状物；*vt.* 给……上蜡；*vi.* 月亮渐满；增大；*adj.* 蜡制的；似蜡的

waxy [ˈwæksi] *adj.* 像蜡的；蜡色的；苍白的

welfare [ˈwɛlˈfɛr] *n.* 福利；幸福；福利事业；安宁；*adj.* 福利的；接受社会救济的

well-being [ˌwɛlˈbiɪŋ] *n.* 幸福；生活安宁；福利；

wheat [wiːt] *n.* 小麦；小麦色

whip [wɪp] *vt.* 鞭打，抽打；严厉地折磨，责打或责备，迫使；把……打起泡沫

wholesomeness [ˈhəʊlsəmnɪs] *n.* 有益健康；正派；健全；生机勃勃；健康向上；有益身心

wrap [ræp] *vt.* 包；缠绕；用……包裹（或包扎，覆盖等）；掩护

wrist [rɪst] *n.* 手腕；腕关节 *vt.* 用腕力移动

Y

yearling [ˈjɪəlɪŋ] *n.* 一岁家畜；满一岁的动物；*adj.* 一岁的

Z

zona pellucida 透明带

zoo [zuː] *n.* 动物园

zoonosis [ˌzuːəˈnəʊsɪs] *n.* 人畜共患病